Dhidik Prastiyanto

Temperature- and Time-Dependent Dielectric Measurements and Modelling on Curing of Polymer Composites

Karlsruher Forschungsberichte aus dem
Institut für Hochleistungsimpuls- und Mikrowellentechnik

Herausgeber: Prof. Dr.-Ing. John Jelonnek

Band 7

Temperature- and Time-Dependent Dielectric Measurements and Modelling on Curing of Polymer Composites

by
Dhidik Prastiyanto

KIT Scientific Publishing

Dissertation, Karlsruher Institut für Technologie (KIT)
Fakultät für Elektrotechnik und Informationstechnik, 2015
Referenten: Prof. Dr. rer. nat. Dr. h.c. Manfred Thumm
Prof. Dr.-Ing. Ellen Ivers-Tiffée

Impressum

KIT Scientific Publishing

Karlsruher Institut für Technologie (KIT)
KIT Scientific Publishing
Straße am Forum 2
D-76131 Karlsruhe

KIT Scientific Publishing is a registered trademark of Karlsruhe Institute of Technology. Reprint using the book cover is not allowed.

www.ksp.kit.edu

Print on Demand 2016

ISSN 2192-2764
ISBN 978-3-7315-0424-5
DOI: 10.5445/KSP/1000048518

Foreword of the Editor

Composite materials and biopolymers gain a lot of interest in industry ranging from avionic, automotive to medical applications. Thermal curing of those materials using conventional heating technologies is energy and time consuming. The use of microwave technology offers selective and volumetric heating that can save energy and speed up the process. In order to simulate, optimize and control microwave assisted curing, the dielectric properties of the processed materials are needed.

Dr.-Ing. Dhidik Prastiyanto developed an in-situ dielectric measurement test-set at 2.45 GHz for monitoring the curing of polymer composites. A non-resonant transmission and reflection method using a vector network analyzer and a standard rectangular waveguide (WR340) measurement unit with conventional heater was selected. The system implements a neural network algorithm to provide fast and reliable reconstruction of the dielectric properties. The neural network is trained and validated with a dataset of scattering parameters and dielectric properties that is generated by a corresponding CST Microwave Studio model in the range of interest.

Careful validation of the dielectric measurements has been carried out by testing well-known materials and by comparing to a numerical reference method. Automatic, computer based temperature-control and data acquisition allows flexible and reproducible investigations of polymer composites curing. The influence of different resins, hardeners and bio-filler on the dielectric and curing behavior has been investigated. The obtained temperature and time dependent dielectric properties of the materials can be used to model the reaction kinetic of the curing process.

As a result, the dielectric behavior along a preset temperature profile as well as the curing status can be predicted with excellent accuracy (better than 96 %).

Temperature- and Time-Dependent Dielectric Measurements and Modelling on Curing of Polymer Composites

Zur Erlangung des akademischen Grades eines

DOKTOR-INGENIEURS

von der Fakultät für
Elektrotechnik und Informationstechnik
des Karlsruher Institut für Technologie (KIT)

genehmigte

DISSERTATION

von

M. Sc. Dhidik Prastiyanto
geb. in Sragen, Indonesien

Tag der mündlichen Prüfung: 11.06.2015

Hauptreferent: Prof. Dr. rer. nat. Dr. h.c. Manfred Thumm
Korreferent: Prof. Dr.-Ing. Ellen Ivers-Tiffée

Kurzfassung

In dieser Arbeit wird ein Test-Set für dielektrische Messungen zur Aushärtung von Polymerkompositen mit Mikrowellen bei 2,45 GHz entwickelt. Design und Simulation werden durchgeführt, um ein optimales Messsystem zur Überwachung der Aushärtung zu erhalten. Eine schnelle Rekonstruktion der dielektrischen Eigenschaften aus den gemessenen *S*-Parametern wird unter Verwendung eines neuronalen Algorithmus erreicht Eine detaillierte Validierung und Verifizierung werden durchgeführt, um genaue und verlässliche Messergebnisse zu erzielen. Die Validierung des aus den dielektrischen Eigenschaften berech-neten Vernetzungsgrades des Polymers wird durch den Vergleich mit einer bestehenden Technik (DSC: Differential Scanning Calorimetry) und durch Messung von gut definierten Materialien durchgeführt.

Für temperatur- und zeitabhängige dielektrische Messungen bei 2.45 GHz wird ein automatisiertes System entwickelt. Der vorgeschlagene experimentelle Aufbau ermöglicht flexible Verfahren zur Überwachung der Aushärtung. Verschiedene Heizungsprofile wie linearer Temperaturanstieg, konstante oder stufenweise erhöhte Temperatur können mit dem entwickelten System, das konventionelle Heizung verwendet, leicht untersucht werden. Die Modellierung von Härtungsverfahren bei 2,45 GHz unter Verwendung von der Dielektrizitätskonstanten und des dielektrischen Verlustfaktors ergibt ein genaueres Modell im Vergleich zur Niederfrequenz-Modellierung, die nur den dielektrischen Verlustfaktor berücksichtigt. Das neue Modell berücksichtigt den Heizvorgang von Raumtemperatur bis zu einer bestimmten spezifischen Härtungstemperatur. Dieses Schema ergibt ein Modell, das näher an der realen Aushärtung ist als das bereits existierende Modell.

Die erhalten Härtungsgrade werden mit den Ergebnissen von DSC-Messungen verglichen. Die Unterschiede zwischen dem vorgeschlagenen Modell und den DSC-Messungen sind 1,5% bzw. 5% für langsame bzw. schnelle Härter. Die Wirkungen verschiedener Härter, unterschiedlicher Mengen an Füllstoff und anderer Temperaturen werden untersucht.

Abstract

In this work a test set for dielectric measurements at 2.45 GHz during curing of polymer composites is developed. Design and simulation are performed to obtain an optimal measurement system for curing monitoring. Fast reconstruction of dielectric properties from measured scattering parameters (*S*-parameters) is solved using a neural network algorithm. Detailed validation and verification are conducted to achieve accurate and realible measurement results. The validation of the curing status, which is derived from the trend of dielectric properties is done by comparing with an existing method (DSC: differential scanning calorimetry) and by measurement of well defined materials.

An automatic system is developed for temperature and time dependent dielectric measurements at 2.45 GHz. The proposed experimental setup allows flexible schemes of curing monitoring. Different profiles of heating such as linear temperature ramp-up, constant or stepwise increased temperature can be easily investigated using the developed system with conventional heating. Modelling of the curing process at 2.45 GHz using both dielectric constant and dielectric loss factor results in a more accurate model compared to low frequency modeling that only uses the dielectric loss factor. The novel model considers the heating process from room temperature up to a specific curing temperature. This scheme provides a model which is closer to the real curing process than the already existing model.

The obtained curing states are compared to the results of DSC. The differences between the model proposed here and the DSC measurements are 1.5% and 5% for slow and fast harderners, respectively. Effects of various harderners, different amount of filler and different temperature are investigated.

Contents

List of figures

List of tables

List of abbreviations

AHEW	amine hydrogen equivalent weigth
DEA	dielectric analysis
DGEBA	diglycidyl ether bisphenol A
DSC	differential scanning calorimetry
D-IPD	diglycidyl ether bisphenol A and isophorone diamine
D-Lar	diglycidyl ether bisphenol A and laromin
D-PEA	diglycidyl ether bisphenol A and polyetheramin
EEW	epoxy equivalent weight
IPDA	isophorone diamane
Lar	laromin
LR	Lichtenecker Rother
LSE	longitudinal section electric
LSM	longitudinal section magnetic
MSE	mean square error
MG	Maxwell Garnett
MLP	multilayer perceptron
MUT	material under test

MWS	microwave studio
PEA	polyetheramin
phr	part per hundred of resin
RBF	radial base function
RMSE	root mean square error
TE	tranverse electric
TM	tranverse magnetic
VNA	vector network analyzer

List of used symbols

a, b	broad and narrow zize of a rectangular wave-guide cross-section
$\vec{B}$	magnetic flux vector
c	speed of light
$\vec{D}$	electric displacement vector
e	Euler's number
$\vec{E}$	electric field vector
f	frequency, activation function, reaction kinetic function
g, h	activation function of hidden neuron
$\vec{H}$	magnetic field vector
j	complex number
$\vec{J}$	current density vector
k_0, k_c	free-space wavenumber, cut-off wavenumber
k_p, k_q	wavenumber at air-filled region, dielectric-filled region
$K_d,\ K_i,\ K_p$	derivative, integral, proportional gain
k_B	Boltzmann constant

L	length
m, n	mode index, exponential factor of reaction kinetic
R	gas constant
S	scattering parameter
t	transmission coefficient, time
T	temperature
T_g	glass transision temperature
V	volume fraction
w	weights of network
x, y, z	Cartesian coordinates
Z	impedance
α	attenuation constant, degree of curing
β	phase constant
γ	complex propagation constant
ε, ε_r	complex permittivity, complex relative permittivity
ε_r', ε_r''	real part of complex relative permittivity, imaginary part of complex relative permittivity (dielectric constant, loss factor)
ε_{rm}', ε_{rm}''	dielectric constant of a mixture, loss factor of a mixture

$\varepsilon_s, \varepsilon_\infty$	static relative permittivity,relative permittivity at infinite frequency
ε_0, μ_0	vacuum permittivity, vacuum permeability
ζ	energy need for dipole orientation
μ, μ_r	complex permeability, complex relative permeability
μ_r', μ_r''	real part of complex relative permeability, imaginary part of complex relative permeability
ρ	charge density
σ	conductivity
τ	relaxation time
τ_0	relaxation time constant
Γ	reflection coeficient
ω	angular frequency
$\boldsymbol{\nabla}$	Nabla operator

1 Introduction

The application of high power microwave technology promises innovative and efficient thermal processes in the fields of food, textile, automotive and aerospace production. The advantages of microwave technology compared to conventional heating methods, such as volumetric and selective heating, make the technology more preferable for materials processing [LFT99, FT04]. In order to develop energy efficient microwave systems and processes, microwave dielectric properties of the processed materials such as thermosetting resins have to be investigated. The lack of temperature dependent dielectric property data of processed materials creates barriers for further development. The temperature dependent dielectric properties of polymer composites and material constituents are rarely found in literature for the temperature and microwave frequency range of interest. Even if such data can be found very often the reported material differs from the specific material of interest. Thus the accurate simulation of processes demand a precise dielectric characterization of all materials involved in the process. In case such materials pass through irreversible changes during the process with consecutive changes of dielectric properties, the knowledge about the time dependence of such effects is essential.

Polymer composites (resin, filler and hardener) gain great interest in current research since the composites have a broad range of applications, such as for packaging, medical, electronic, automotive, avionic and engineering applications. The focus of the current research is to find composite materials with certain characteristics that cannot be achieved in single component materials. The advantages, such as enhanced mechanical strength, lighter weight and higher temperature resistance can be obtained by material cross-linking. In order to process the materials, usually conventional heating is applied. Heating the materials with conventional procedures needs much energy and time. Therefore, an alternative technology that offers energy efficient materials processing is necessary. The microwave assisted processing of polymer composites offers these advantages. The development of a microwave system for the processing of polymer composites requires design and simulation of the system.

The curing process of some polymer composites requires constant temperature or stepwise temperature changes for certain time to ensure that the composites are well cured. The curing behavior of the composite depends on its constituents and the applied temperature profile. In addition, different mixtures reveal different absorption of microwave energy so that the penetration depth of microwaves into the mixture needs to be investigated before applying microwave heating. The knowledge of dielectric properties of mixtures during curing is useful for a model predictive control of efficient curing processes with microwaves [SKL+14].

Furthermore for monitoring the curing of thermosetting resins, beside other techniques, dielectric analysis (DEA) can be implemented. The DEA has been utilized for online monitoring curing processes at low frequency around 1 kHz [RBF05, HJPK13]. The technique might be implemented also for microwave frequencies. Then, the development of the dielectric analysis of curing at 2.45 GHz becomes very important.

1.1 Motivation

The 2.45 GHz Industrial, Scientific, Medical (ISM) frequency band has been successfully developed for materials processing at Karlsruhe Institute of Technology (KIT). The microwave heating system called High Electromagnetic Power Heating Autoclave

Injected Structures Oven System (HEPHAISTOS) has been developed to increase the homogeneity of microwave heating. The HEPHAISTOS technology utilizing a hexagonal shape of the processing chamber offers a more homogenous field distribution compared to the other applicator shapes [FT04]. HEPHAISTOS VHM 180/300 with 1.80 m x 1.55 m x 3.30 m of width, height and depth has been developed and implemented in some materials processing applications. Polymer composites curing and other material processing has been successfully demonstrated using this microwave assisted system [Lin+11]. For better understanding of the interaction between microwave energy and the material, the information on the dielectric properties of the materials is essential.

Further development of polymer research is currently focusing on biopolymers. Biopolymers are polymers which are made from living organism or other renewable resources [TKAS12]. Biopolymers gain a lot of interest because the materials can be decomposed naturally. The wide application of biopolymers, ranging from packaging to medical apparatus, also makes the materials very interesting to be investigated. Biopolymers also offer the advantage of low cost of material resources and production systems. Processing biopolymers with microwaves saves energy due to the advantages offered by volumetric heating. In order to see the possibility of processing biopolymers with microwaves, the influence of biofillers on the curing of thermosetting resins has been examined.

The curing behavior of a mixture depends on several factors, such as curing temperature, type of hardener and filler. The best temperature and time scenario of curing can be derived from the changes of the dielectric properties at 2.45 GHz of the mixture during curing.

The changes of dielectric properties during curing depend on the applied temperature profile, type of hardener and amount of filler. In order to investigate the dielectric properties of some types of epoxy resins, hardeners and biofillers, a reliable test set for dielectric measurement during the curing process has to be developed.

1.2 State of the art

Reconstruction of the dielectric properties of several materials has been investigated with various methods in several publications [NR70, SM78, CCFD03, Has10 and AT13]. There is no general method that is suitable for all materials. The methods can be classified into resonant and non-resonant methods with some advantages and disadvantages of each method. As it is well known, the resonant method, such as the cavity perturbation method, is suitable for characterization of low loss materials (dielectric loss tangent less than 1) [CON+04], while the non-resonant method using the measurement of transmission and reflection is suitable for medium-to-high dielectric loss materials [KW98]. The transmission and reflection method is commonly implemented using waveguide, coaxial line or free space measurements. The waveguide based transmission-reflection method has some variations depending on sample shape and placement. Fully and partially covered waveguide cross sections are commonly used for introducing the sample [AFT09]. A simple analytical method of fully covered transmission lines has been published in [NR70] [SM78]. The development of this method can be found in [AFT 06]. There an optimization method was applied for reconstruction of dielectric properties of a fully covering sample. A nonlinear-least square optimization was implemented for minimizing the difference of the measured and calculated scattering parameters (S-parameters). The implementation of this method for monitoring curing will require a crucible that fully covers the waveguide cross-section. This might cause a problem of homogeneity of heat transfer, and also needs a large sample volume. Therefore a partially covered sample is more preferable for measurement of curing. The partially covered sample method can be found in [CCFD03, AFT07 and Has10]. Those are implemented for solid materials only.

Measurements on liquid materials in a small rectangular container had been published in [AFT09]. The dielectric properties were reconstructed using an analytical method. Another analytical method for measurements on powder materials in a glass tube is presented in [AT13]. There a simplification of the glass tube container has been used. An analytical approach for the reconstruction of dielectric properties was implemented. However, the method is only valid for liquids with a relatively low dielectric constant and loss factor because of simplification that ignores high order evanescent modes. Dielectric properties were calculated by a derivation formula for the Longitudinal Section Electric mode (LSE) mode [AFT07]. In order to simplify the problem, the analytical solution only considered the fundamental mode. In order to increase the accuracy of the measurement, the influence of higher order evanescent modes should be considered. The effect of higher order evanescent modes increases for higher dielectric constants and loss factors [QG96]. A further requirement is that the method should be considerably fast for online calculation during curing. Higher order modes can be considered in the simulation with a full 3D electromagnetic modeling software [Cst]. An optimization routine using the results of the 3D simulation then can be a solution for the reconstruction of dielectric properties of thermosetting resins.

The measurement of dielectric properties based on Finite Differential Time Domain (FDTD) simulation [Qui] has been developed in [EMY07]. The concept makes use of a radial base function (RBF) neural network for the reconstruction of dielectric properties. Some liquids were measured and tested at 915 MHz, showing good performance compared to literature data. This fact indicates that a neural network approach might be used for real time monitoring of dielectric properties. Therefore, the development and implementation of a neural network for monitoring the dielectric properties during curing of thermosetting resins has been initiated in the present work. In [EMY07], the dielectric measurement system was developed only for measurement at room temperature. In order to implement the neural network for online curing monitoring, many practical aspects such as the optimal size of crucible, the heating rate during curing and the temperature scenario need to be considered based on the experimental results. The neural network has also to be optimized for characterization of polymer composites. In addition, the validation of the neural network has to be done with standard materials.

The dielectric properties of mixture resins had been studied at 2.45 GHz by using a commercial dielectric probe HP 85070 [Yar01]. The dielectric properties were only measured at constant temperature before and after the curing process. Determination of the dielectric properties during the curing process had been reported in [KC99] [HJPK13]. The system was based on the dielectric analysis of capacitance probe measurements. In [KC99] the dielectric analysis for noninvasive monitoring of the curing process had been developed. The system was based on a coaxial resonator and was only designed for real time monitoring of curing at low frequencies up to 100 kHz. In [HJPK13] the dielectric properties of thermosetting resins had been measured at low frequencies (up to 10 kHz) during the curing process. The development of a measurement system in the present work that is able to monitor the curing process at 2.45 GHz is very important to provide the information about dielectric properties of curing at this ISM microwave frequency.

Measurements of dielectric properties during curing at microwave frequencies have been performed in [MKB92 and NND04] using the cavity perturbation method. The cavity perturbation method is sensitive to the changing of the coupling coefficient that occurs during curing. The shrinkage of the volume during curing decreases the accuracy of the measurement by the cavity perturbation method. Noisy results, with 10% uncertainties of dielectric measurements due to the coupling problem and volume shrinkage during curing can be found in [NND04]. The coupling coefficient depends on changes of dielectric properties during curing. In addition the changing of the volume during curing also gives an additional error. In order to solve these problems, in the present work a non-resonant transmission and reflection method is developed. The method has no issue with the coupling coefficient problem and is less sensitive to volume shrinkage.

A loss factor model during curing has been published in [KC99 and HJPK13]. The model shows good agreement compared with DSC and Raman spectroscopy. However, the model is only valid for isothermal curing. In the real experiment, it is impossible to achieve isothermal curing starting from time zero and room temperature to a specific curing temperature. Preheating the sample surely initiates the cross-linking reaction. In order to obtain a more accurate curing model, a model that considers the heating stages to the specific curing temperature has to be developed.

1.3 Objectives of the study

The main goal of this work is to develop an accurate and reliable dielectric measurement system for curing. The proposed system is designed and realized for reconstruction of temperature dependent dielectric properties at 2.45 GHz. The setup is implemented and optimized with focus on polymer composites. Different composite compounds consisting of resin, filler and hardener can be measured by using the system to obtain the temperature dependent dielectric properties. Not only dielectric properties of separated compounds, but also the dielectric properties of mixtures during curing are investigated. Temperature and time dependent dielectric properties of the curing process are measured by the proposed technique. Furthermore, mixing rule formulas for dielectric mixtures are studied. Curing of thermosetting resins with different curing temperatures, hardeners, amount of filler are carried out to investigate the effect of these parameters on the curing behavior. Modeling of the temperature and time dependent dielectric properties is performed for each mixture. A novel model of the curing process using both dielectric constant and dielectric loss factor is developed in this study. The other new investigation which has not been published in literature yet is the implementation of a temperature and time scheme of the curing process as close as possible to the real curing. Both measurement and modeling are performed by implementing a slow heating process starting from room temperature up to some specific curing temperatures.

In order to measure dielectric properties, the material under test (MUT) is inserted in a standard waveguide WR340 (Waveguide, Rectangular 340). In this study, the transmission and reflection method with sample geometries that partially cover the waveguide cross-section is selected. Since the dielectric properties are measured during the curing process, the procedure starts by measuring the liquid mixture (resin, hardener and filler) and it is followed by measuring the phase transition from liquid to solid. The third procedure is the measurement of the cured materials which are finally in solid form. As already mentioned, the partially covered waveguide method is appropriate for measurement of liquid and solid materials. Dealing with the heating process of the sample, a smallest size of the sample with acceptable accuracy and precision is required. Small sample is needed to ensure uniform heat distribution. In order to reconstruct dielectric properties from measured scattering parameters (*S*-parameters), a neural network algorithm is applied. Detailed analyses of error sources are performed to get a well-defined accuracy of the measurement system. After all sources of errors have been investigated, the estimation of the total error of the characterization and of dielectric properties is calculated.

The proposed method considers both dielectric constant and dielectric loss factor for monitoring of curing. The proposed setup is designed as an integrated heating and automatic control system for automatic scheduling of temperature and time dependence of the dielectric measurement during curing. Different curing scenario can be evaluated to obtain optimal curing. Ramp, constant, and combination schemes can be flexibly investigated using the proposed system. More realistic dielectric data and models during curing can be obtained with this approach. In this work, the model of temperature dependent dielectric properties during heating is represented by the Debye model [Deb29] and sigmoidal model [JB05]. The mixing rule formula is investigated for mixtures of resin, hardeners and fillers. Then, the novel model of curing is developed by the combination of the temperature dependent model for pure materials with the curing model. The models are compared with the measurement results. The work starts with the simulation of the measurement system to obtain an appropriate measurement setup. Defining the crucible for the sample is an important step to obtain the optimal dimension, in order to achieve the necessary accuracy and precision of the measurement result. The second step is the development of a fast dielectric reconstruction from measured *S*-parameters with a neural network algorithm. A training and testing dataset has to be defined appropriately considering the range of dielectric properties of the sample. The dataset is generated from a 3D model of the selected measurement setup. Then, the neural network is trained with the generated dataset. The network is verified using some other datasets and validated with measurements on well-defined materials. The third step is the implementation of the proposed technique for an automatic measurement of dielectric properties during curing. Automatic temperature control and data acquisition is developed to achieve a reliable and reproducible system for monitoring of curing. The last step is modelling the obtained dielectric property data.

1.4 Structure of this work

The introduction, motivation, state of the art and the main objectives of the research are explained in Chapter 1. Then, Chapter 2 describes the theory of dielectric properties, dielectric measurement methods and the transmission-reflection technique. The method and simulation model used are explained in Chapter 3. Experimental setup, validation and materials are described in Chapter 4. Further, Chapter 5 presents the results of dielectric measurements. Temperature dependent measurements on the materials are presented. The measurement of curing is explained also in Chapter 5. Chapter 6 is dedicated to modeling of the obtained dielectric data and their discussion. The novel model of curing is explained and verified with the experimental results. Conclusion and outlook are presented in Chapter 7.

2 Theory

In this Chapter, the theory of dielectric properties and measurement techniques are described. First, the Maxwell equations are listed followed by the theory of polarization mechanisms. The dielectric measurement methods are explained in a general overview. Then the theory of the transmission and reflection method is described.

2.1 Theory of dielectric properties

2.1.1 Maxwell equations

The relation between the electric and magnetic field is described by the Maxwell equations. Considering time harmonic fields according to $e^{j\omega t}$ where ω is the angular frequency and $\frac{\partial}{\partial t} = j\omega$, the Maxwell equations are written as [Col91, Poz12]:

$$\nabla \times \vec{H} = j\omega\vec{D} + \vec{J} \tag{2.1}$$

$$\nabla \times \vec{E} = -j\omega\vec{B} \tag{2.2}$$

$$\nabla . \vec{D} = \rho \tag{2.3}$$

$$\nabla . \vec{B} = 0 \tag{2.4}$$

where $\vec{E}$, $\vec{H}$, $\vec{D}$ and $\vec{B}$ are the electric field, the magnetic field, the electric displacement and the magnetic flux, respectively. The current density vector and the charge density are $\vec{J}$ and ρ, respectively.

The relations between the electromagnetic fields in homogeneous materials are described by the following constitutive equations:

$$\vec{D} = \varepsilon \vec{E} \tag{2.5}$$

$$\vec{B} = \mu \vec{H} \tag{2.6}$$

$$\vec{J} = \sigma \vec{E} \tag{2.7}$$

Permittivity (ε), permeability (μ) and conductivity (σ) determine how the electromagnetic field changes in materials. Based on their conductivity, materials can be classified as insulators, semiconductors and conductors [CON+04]. In this study only materials with very low conductivity (dielectric materials) are investigated. The relative complex permittivity ε_r is described by the following equation:

$$\varepsilon_r = \frac{\varepsilon}{\varepsilon_0} = \frac{\varepsilon' - j\varepsilon''}{\varepsilon_0} = \varepsilon_r' - j\varepsilon_r'' \tag{2.8}$$

Here $\varepsilon_0 = 8.854 \text{ x } 10^{-12}$ F/m is the permittivity of free space. The real and imaginary parts of the relative complex permittivity are described by ε_r' (relative permittivity or dielectric constant) and ε_r'' (loss factor), respectively.

The relative complex permeability μ_r is described by the formula:

$$\mu_r = \frac{\mu}{\mu_0} = \frac{\mu' - j\mu''}{\mu_0} = \mu_r' - j\mu_r'' \tag{2.9}$$

Here $\mu_0 = 4\pi \text{ x } 10^{-7}$ H/m is the permeability of free space. The real and imaginary parts of the relative complex permeability are described by μ_r' and μ_r'', respectively. Since thermosetting resins, hardeners and bio fillers are non-magnetic materials, the relative complex permeability is assumed to be equal to $\mu_r = 1$.

2.1.2 Dielectric polarization

Polarization and loss mechanisms in materials strongly depend on material composition as well as frequency and temperature range and can be classified into electronic or ionic conduction, dipolar relaxation and atomic or electronic polarization [CON+04]. For dielectric materials where conductive losses can be neglected in the microwave frequency range, the dielectric properties are mainly influenced by dipolar relaxation. In the frequency range above the infrared region, the permittivity is influenced by atomic or electronic polarization. Since the present work is focused on microwave frequencies, only dipolar polarization is explained further which can be described with the Debye model [Deb29].

The formula of the Debye model for the complex relative permittivity is given by:

$$\varepsilon_r = \varepsilon_\infty + \frac{\varepsilon_s - \varepsilon_\infty}{1 - j\omega\tau} \tag{2.10}$$

where:

ε_∞ = relative permittivity at infinite frequency

ε_s = static relative permittivity

τ = relaxation time in s

ω = angular frequency in Hz.

The Debye model can be separated into the real and the imaginary part of the relative complex permittivity by multiplying with the complex conjugate:

$$\varepsilon_r' = \varepsilon_\infty + \frac{\varepsilon_s - \varepsilon_\infty}{1 + (\omega\tau)^2} \tag{2.11}$$

$$\varepsilon_r'' = \frac{\omega\tau(\varepsilon_S - \varepsilon_\infty)}{1 + (\omega\tau)^2} \tag{2.12}$$

The relaxation time is a function of temperature typically following an Arrhenius equation as follows [ZZWD07]:

$$\tau = \tau_0 \exp(\frac{\zeta}{k_B T}) \tag{2.13}$$

with:

τ_0 = relaxation time constant in s

ζ = activation energy needed for dipole orientation in J/mol

k_B = Boltzman constant 1.38E-23 J/K

T = temperature in K.

The Debye parameters depend on the material characteristics. As can be seen from equations 2.11 - 2.13, both dielectric constant and loss factor are frequency and temperature dependent. Experimental results on frequency and temperature dependent loss factor for some samples can be found in [Suc10].

As an example, the temperature dependent dielectric constant of the resin diglycidyl ether bisphenol A (DGEBA) with $\varepsilon_\infty = 3.6$ and $\varepsilon_s = 7.0$ according to the Debye model for different τ_0 at f = 2.45 GHz and $\zeta = 3.65\text{E-}20$ J/mol is presented in Fig. 2.1.

The corresponding loss factor is shown in Fig. 2.2. Following equation 2.12, the loss factor will be maximum if $\omega\tau = 1$. Therefore at identical frequency with smaller τ_0 the maximum of the loss factor appears at higher temperatures as shown in Fig. 2.2. Here the variable τ_0 is variated in order to show the effect of different relaxation time. The detailed investigation is presented in Chapter 6.

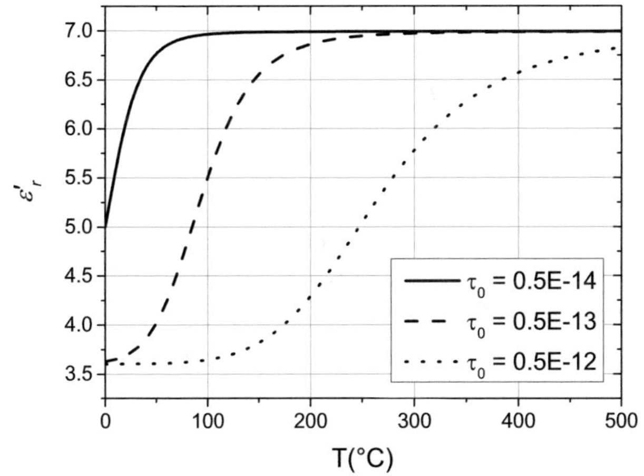

Figure 2.1: Example of Debye model of temperature dependent dielectric constant

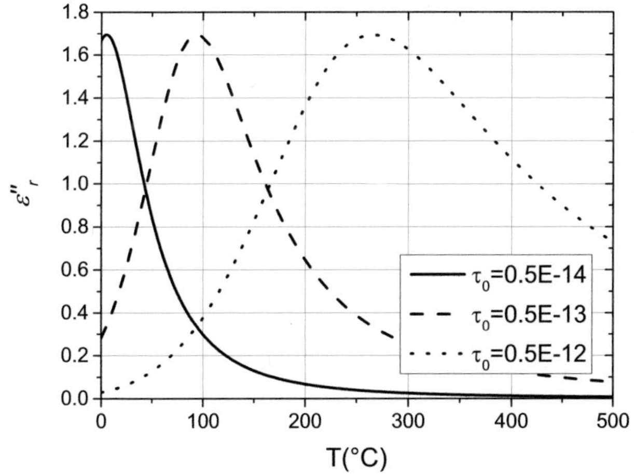

Figure 2.2: Example of Debye model of temperature dependent loss factor

2.1.3 Dielectric mixing rules

Dielectric properties of a material mixture can be derived from the dielectric properties of each constituent. The dielectric constant of a mixture can be calculated with two common mixing rules which are the Lichteneker and Rothers (LR) formula [ZLV98, QB12] and the Maxwell Garnett (MG) formula [Sih00, QB12]. The MG formula is used for heterogenous mixtures with small volume fraction of filler in the host matrix [BB03]. The LR formula can be wider applied than the MG formula from anisotropic to isotropic mixtures. The LR formula is described as:

$$\varepsilon'_{rm} = \left(\sum_{i=1}^{N} V_i \varepsilon'^{\varsigma}_{ri}\right)^{1/\varsigma} \tag{2.14}$$

The variable V_i is the relative volume fraction of the constituent i, where i = 1, 2,..., N, so that the sum of all fractions is equal to 1. The exponential variable ς describes a specific microgeometry (topology) of a composite [GLV00] and varies in the range of [-1,1] where $\varsigma \neq 0$ [ZLV98, QB12]. If the dielectric properties of each component and the volume fraction are known, the dielectric properties of the mixture can be calculated from formula 2.14. The correct value of dielectric properties can be determined by selecting the correct value of the exponential factor ς. The determination of ς is based on experimental results usually.

As an example, the mixing for DGEBA and the hardener Polytheramin (PEA) with $\varepsilon'_{r1} = 3.73$ and $\varepsilon'_{r2} = 4.56$, and the volume fraction $V_1 = 0.72$ and $V_2 = 0.28$ respectively is investigated. The estimated dielectric constant of the mixture is presented in Fig. 2.3 as a function of the exponential factor ς.

The MG formula is given by:

$$\frac{\varepsilon'_{rm}-\varepsilon'_{r1}}{\varepsilon'_{rm}+2\varepsilon'_{r1}} = V_2 \frac{\varepsilon'_{r2}-\varepsilon'_{r1}}{\varepsilon'_{r2}+2\varepsilon'_{r1}} \tag{2.15}$$

Using the same mixture as for the LR formula, the calculated dielectric constant is larger: $\varepsilon'_{rm} = 4.05$.

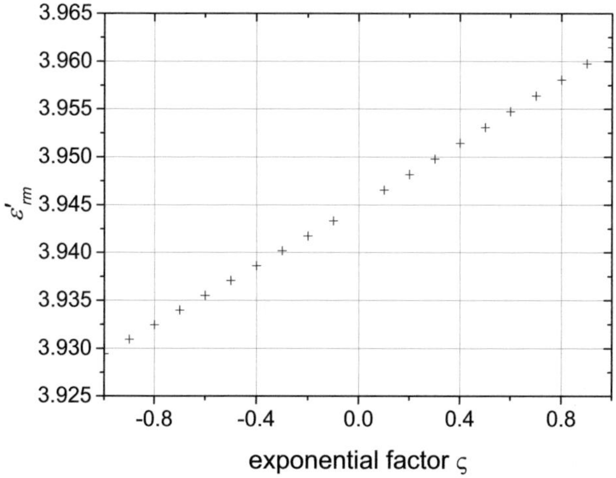

Figure 2.3: Dependence of the dielectric constant of a mixture on ς calculated with the LR formula for mixing of DGEBA and PEA with $V_1 = 0.72$ and $V_2 = 0.28$

The formulas 2.14 and 2.15 can be applied also for the calculation of the loss factor by simply replacing the dielectric constants by the loss factors. Detailed investigation on the LR and MG formulas is presented in Chapter 6.

2.2 Dielectric measurement methods

Dielectric measurement methods can be classified into resonant and non-resonat methods. The resonant method is based on the theory of microwave resonance, while the non-resonant method is based on microwave propagation.

2.2.1 Resonant methods

In [CON+04], resonant methods are categorized as resonator and resonant-pertubation methods. The resonator method uses the sample as pertubation of the resonator. Dielectric properties of the sample are derived from the resonant frequency and the quality factor of the dielectric resonator. A common dielectric resonator is described in [Cou70].

In the resonant-pertubation method, dielectric properties are calculated based on pertubation theory. Introducing a sample to a resonator with given boundary condition will lead to a change of resonant frequency and quality factor. In this resonant-pertubation method in order to attain accurate measurement results, a small sample is placed in a pure electric field area. The dielectric properties are often calculated using the cavity perturbation formulas that can be found in [KN92, CON+04].

2.2.2 Non resonant methods

Non resonant methods use impedance and wave phase velocity in the materials for dielectric properties characterization. The dielectric properties can be calculated from the reflection of the electromagnetic wave from the sample which is known as reflection method. Another method is known as transmission and reflection method. Here the dielectric properties are calculated from the transmission and reflection coefficients of the electromagnetic wave that propagates from one port to another port through the material [Jav90]. If the lengths of the waveguide section before and behind the sample are sufficiently long so that evanescent modes can be neglected, the position of the sample along the direction of wave propagation has no influence on the sensitivity of the measurement. As long as the exact position is known, the position of the sample along the direction of wave propagation can be included in the phase calculation for the reconstruction of the dielectric properties. However, the position of the sample in transversal direction of wave propagation influences the sensitivity of this measurement method.

2.3 Transmission and reflection method

2.3.1 Fully covered waveguide cross section

The transmission-reflection method uses both the transmission and reflection coefficients to calculate the properties of materials. A sample can be inserted into a segment of a transmission line such as a waveguide or a coaxial line. A sample loaded waveguide section is shown in Fig. 2.4 [Jav90]. An incident electromagnetic wave is partially reflected from and transmitted through the sample.

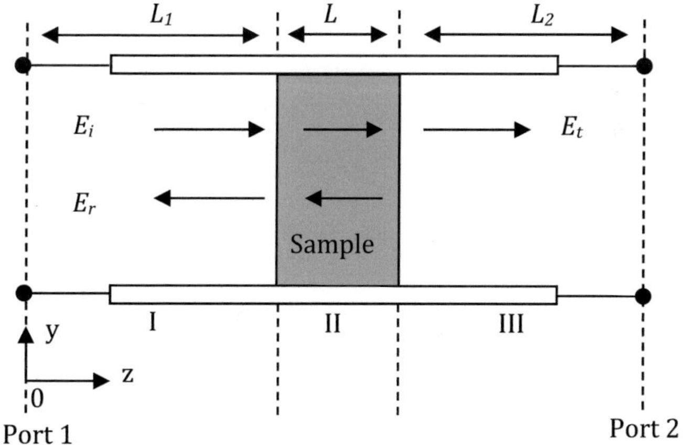

Figure 2.4: Transmission line with inserted sample

The corresponding fractions depend on the characteristics of the material. If E_I, E_{II} and E_{III} are the normalized electric fields of region I, II and III, respectively, the spatial distributions of these fields are:

$$E_I = e^{-j\beta_0 z} + C_1 e^{j\beta_0 z} \tag{2.16}$$

$$E_{II} = C_2 e^{-\gamma z} + C_3 e^{\gamma z} \tag{2.17}$$

$$E_{III} = C_4 e^{-j\beta_0 z} \tag{2.18}$$

where γ and β_0 are the propagation constants in the dielectric (II) and air filled regions I and III, respectively, with $\gamma = \alpha + j\beta$, where α is the attenuation constant and β the phase constant in the sample.

The constants C_1 and C_2 are determined from the boundary conditions of electric and magnetic fields at $z = L_1$, and C_3 and C_4 at $z = L_1 + L$. The boundary condition for the transverse electric field is the continuity of the tangential component at the interfaces, while the boundary condition for the magnetic field is that its tangential component is continuous across the interfaces with the assumption that no surface current is generated [Jav90]:

$$E_I\big|_{z=L_1} = E_{II}\big|_{z=L_1} \tag{2.19}$$

$$E_{II}\big|_{z=L_1+L} = E_{III}\big|_{z=L_1+L} \tag{2.20}$$

$$\frac{1}{\mu_0}\frac{\partial E_I}{\partial z}\bigg|_{z=L1} = \frac{1}{\mu_0\mu_r}\frac{\partial E_{II}}{\partial z}\bigg|_{z=L1} \tag{2.21}$$

$$\frac{1}{\mu_0\mu_r}\frac{\partial E_{II}}{\partial z}\bigg|_{z=L_1+L} = \frac{1}{\mu_0}\frac{\partial E_{III}}{\partial z}\bigg|_{z=L_1+L} \tag{2.22}$$

The scattering parameters of the two port network can be obtained by solving equation 2.16-2.18 with the above boundary conditions 2.19-2.22. As given in [Jav90, NR70], the scattering parameters at the interfaces of the sample are:

$$S_{11} = \left[\frac{\Gamma(1-t^2)}{1-\Gamma^2 t^2}\right] \tag{2.23}$$

$$S_{22} = \left[\frac{\Gamma(1-t^2)}{1-\Gamma^2 t^2}\right] \tag{2.24}$$

$$S_{21} = \left[\frac{t(1-\Gamma^2)}{1-\Gamma^2 t^2}\right] \tag{2.25}$$

where the transmission coefficient of the sample is defined as:

$$t = e^{-\gamma L} \tag{2.26}$$

while the reflection coefficient is given by:

$$\Gamma = \frac{Z-Z_0}{Z+Z_0} = \frac{\beta_0-\gamma}{\beta_0+\gamma} \tag{2.27}$$

where the impedances in the air filled and dielectric regions are $Z_0 = \frac{\omega\mu_0}{\beta_0}$ and $Z = \frac{\omega\mu_0}{\gamma}$, respectively. In the present work, a rectangular waveguide WR340 is selected as sample holder. The phase constants in the sample and in air can be calculated from the following formula [Poz12]:

$$\beta = \sqrt{k^2 - k_c^2} \tag{2.28}$$

where k is the wavenumber and k_c is the cutoff wavenumber. The wavenumber and cutoff wavenumber in rectangular waveguides can be calculated with the following formulas:

$$k = \omega\sqrt{\mu^*\varepsilon^*} \tag{2.29}$$

$$k_{c_{m,n}} = \sqrt{(m\pi/a)^2 + (n\pi/b)^2} \tag{2.30}$$

where m and n are mode numbers and a and b are the broad and narrow inner side lengths of the waveguide, respectively. Assuming that only the fundamental TE_{10} mode (tranverse electric mode) exists, the cutoff wavenumber is given by:

$$k_{c_{10}} = \pi/a \tag{2.31}$$

Then the propagation constant (phase constant) in the air filled part of the waveguide can be written as:

$$\beta_0 = \sqrt{\omega^2 \varepsilon_0 \mu_0 - (\frac{\pi}{a})^2} \tag{2.32}$$

while the phase constant in the sample filled part of the waveguide can be calculated with:

$$\beta = \sqrt{\omega^2 \varepsilon_r^* \varepsilon_0 \mu_r^* \mu_0 - (\frac{\pi}{a})^2} \tag{2.33}$$

An analytical method for the calculation of dielectric properties of a sample that fully covers the waveguide cross section can be found in [NR70 and Wei74]. For improvement of the analytical method, numerical calculations have been applied for searching the optimal values of the dielectric properties of the material iteratively [JVK90, AFT06]. The main idea of optimization is comparing measured and calculated S-parameters of a simple measurement setup. The optimization algorithm can be described in the following steps:

1. Definition of an initial value of dielectric constant and loss factor. A good choice of the initial values of these parameters decides the convergence of the optimization.
2. Perform an optimization procedure with the following functions:
   - Measure S-parameter data: S_{11m} and S_{21m}.

- Define sample thickness d and distances L_1 and L_2 of the sample interfaces to port 1 and port 2, respectively.
- Transform the measured S-parameters at the reference planes (port 1 and 2) to the faces of the sample: $S_{11} = S_{11m}e^{2j\beta_0 L_1}$ and $S_{21} = S_{21m}e^{j\beta_0(L_1+L_2)}$.
- Calculate phase constant in air and dielectric media with formulas 2.32 and 2.33, respectively.
- Calculate the transmission and reflection coefficients of the first and the second interface based on the formulas 2.26 and 2.27.
- Minimize the error function of measured and calculated S-parameters (2.23 and 2.24).
- Search for optimal values of dielectric constant and loss factor based on minimal error function as initial values for the next iteration step.

3. Do the optimization procedure for the new initial values.
4. Select the optimal values via the minimal value of the error function.

2.3.2 Partially covered waveguide cross section

Analysis of partially filled waveguides has been explained in [AFT07, CCFD03, Chu06, Col91 Has 93 and Poz12]. TE and TM modes cannot satisfy the boundary conditions in the dielectric region individually. Instead hybrid modes are solutions and satisfy the boundary conditions. These modes are the longitudinal section electric (LSE) modes (distorted TE modes) and the longitudinal section magnetic (LSM) modes (distorted TM modes). The characteristic equation or eigenvalue equation is obtained using the transverse resonance method.

In Fig. 2.5, the transmission line cross-section with symmetrically placed partially covering sample and the corresponding equivalent circuit are shown. In the equivalent circuit, the impedance Z_0

in the air filled parts and Z in the dielectric sample are inversely proportional to the correspoding wavenumbers k_p and k_q, respectively.

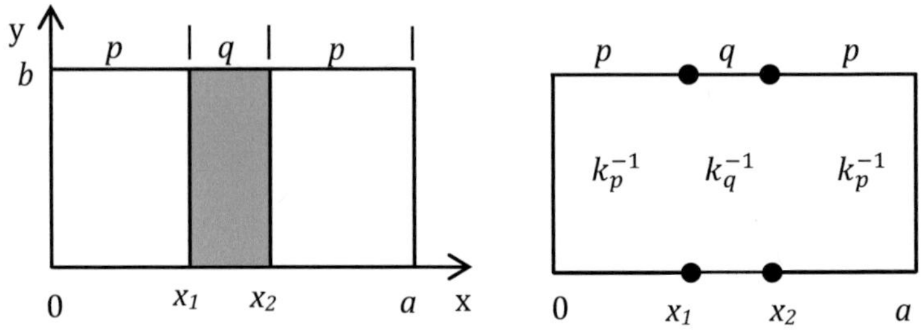

Figure 2.5: Waveguide cross section partially filled with a symmetrically material sample (left) and equivalent circuit (right)

The transverse resonance method uses the fact that at any tranverse plane in the line, the sum of the input impedance seen looking to the right and left side must be zero [Poz12, Col91]:

$$Z_{in}^{r}(x) + Z_{in}^{l}(x) = 0 \quad \text{for all } x \tag{2.34}$$

In a lossless transmission line with impedance Z_0, wavenumber β_0 and terminated load Z_L, the input impedance at the distance l can be calculated by [Poz12]:

$$Z_{in} = Z_0 \frac{Z_L + jZ_0 \tan \beta_0 l}{Z_0 + jZ_L \tan \beta_0 l} \tag{2.35}$$

If Z_L=0 (short circuit), the input impedance will be:

$$Z_{in} = jZ_0 tan\beta_0 l \tag{2.36}$$

If Z_L=∞ (open circuit), the input impedance is given by:

$$Z_{in} = -jZ_0 \cot \beta_0 l \tag{2.37}$$

Using equations 2.34, 2.35 and 2.36, the equivalent circuit in Fig. 2.5 can be analyzed. The input impedance at x = x_2 (looking to the right) can be written as:

$$Z_{in}|_{x=x_2} = jZ_0 \, tan \, k_p p \tag{2.38}$$

The input impedance at x = x_1 (looking to the right) can be calculated from formula 2.35 where $Z_{in}|_{x=x_2}$ is considered as a load:

$$Z_{in}|_{x=x_1} = Z \frac{Z_{in}|_{x=x_2} + jZ \, tan \, k_q q}{Z + jZ_{in}|_{x=x_2} \, tan \, k_q q} \tag{2.39}$$

Insertion of equation 2.38 into 2.39 gives:

$$Z_{in}|_{x=x_1} = Z \frac{jZ_0 \, tan \, k_p p + jZ \, tan \, k_q q}{Z - Z_0 \, tan \, k_p p \, tan \, k_q q} \tag{2.40}$$

On the other hand, the input impedance at x = x_1 (looking to the left) is described as:

$$Z_{in}|_{x=x_1} = jZ_0 \, tan \, k_p p \tag{2.41}$$

Based on equation 2.34 and considering that the impedance is inversely proportional to the wavenumber, equations 2.40 and 2.41 can be combined to:

$$-\mathrm{j} \, {k_p}^{-1} \tan k_p p = \frac{1}{kq} \frac{\mathrm{j}{k_p}^{-1} \tan k_p p + \mathrm{j}{k_q}^{-1} \tan k_q q}{{k_q}^{-1} - {k_p}^{-1} \tan k_p p \, \tan k_q q} \tag{2.42}$$

Some simplifications lead to the following eigenvalue equation:

$${k_p}^2 \tan k_q q + 2k_p k_q \tan k_p p - {k_q}^2 \tan^2 k_p p \tan k_q q = 0 \tag{2.43}$$

Since the propagation along the guide must be the same in the empty portion and dielectric-filled portion of the rectangular waveguide, the propagation constant for nonmagnetic materials is given by [Col91, CCFD03]:

$$\gamma^2 = k_q{}^2 + (\frac{n\pi}{b})^2 - \varepsilon_r k_0{}^2 = k_p{}^2 + (\frac{n\pi}{b})^2 - k_0{}^2 \qquad (2.44)$$

A simple transverse resonance condition for a symmetrically placed dielectric slab in the rectangular waveguide is given in [Col91]. The transverse resonance condition is equivalent to short circuit and open circuit at the center of the slab for asymmetrical modes and symmetrical modes, respectively.

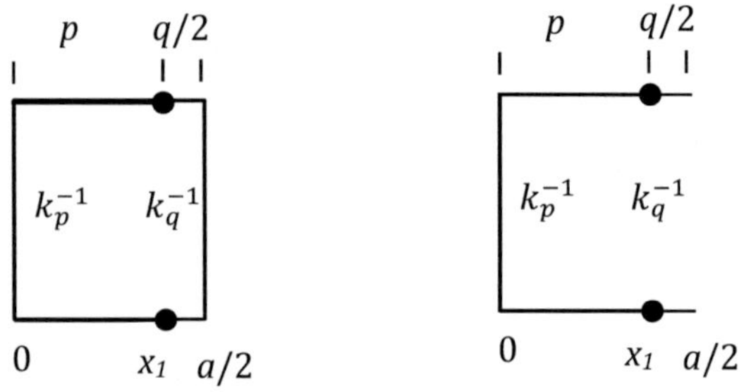

Figure 2.6: Tranverse equivalent circuit for a symmetrically placed dielectric slab in rectangular waveguide for asymmetrical modes (left) and symmetrical modes (right)

The input impedance at x = x_1 (looking to the left) for both asymmetrical and symmetrical modes is given by equation 2.41. The input impedance at x = x_1 (looking to the right) for asymmetrical modes is like in equation 2.36:

$$Z_{in}|_{x=x_1} = jZ\, tan\frac{k_q q}{2} \qquad (2.45)$$

Using equation 2.34, the eigenvalue of asymmetrical modes can be written as:

$$jk_p{}^{-1} \tan k_p p = -jk_q{}^{-1} \tan\frac{k_q q}{2} \qquad (2.46)$$

which can be simplified to:

$$k_q \tan k_p p = -k_p \tan \frac{k_q q}{2} \tag{2.47}$$

The input impedance at x = x_1 (looking to the right) for symmetrical modes is like in equation 2.37 (open circuit):

$$\mathrm{Z}_{in}|_{x=x_1} = -jZ \cot \frac{k_q q}{2} \tag{2.48}$$

Using equation 2.34, equations 2.41 and 2.48 can be combined to:

$$jk_p{}^{-1} \tan k_p p = jk_q{}^{-1} \cot \frac{k_q q}{2} \tag{2.49}$$

A simplification based on trigonometric properties leads to:

$$k_p \cot k_p p = k_q \tan \frac{k_q q}{2} \tag{2.50}$$

The eigenvalue equations 2.47 and 2.50 are the same as the corresponding equations in [Col91, CCFD03, Has09]. Only symmetrical modes (equation 2.50) were considered for dielectric properties reconstruction in [AFT07, CCFD03 and Chu06].

If the dielectric slab is thin enough compared to the broad size of waveguide cross-section ($q \leq 0.25a$), the effect of higher order modes can be neglected [CCFD03, Has03 and AFT 07]. Therefore in these publications only the LSE_{10} mode is considered. Also in the present work the size of the sample in the Teflon crucible is selected to fulfill this requirement. Detailed simulation of the Teflon crucible is described in Chapter 3.

Also in the present work, considering symmetrical modes, formula 2.44 together with equation 2.50 is used for the calculation of dielectric properties with the assumption that only the LSE_{10} mode exists. Considering the propagation constant $\gamma = \alpha + j\beta$ and the wavenumber $k_q = k'_q + jk''_q$ in the dielectric sample, the complex relative permittivity can be obtained from formula 2.44:

$$(\alpha + j\beta)^2 = (k'_q + jk''_q)^2 - \varepsilon_r {k_0}^2 \tag{2.51}$$

Reformulation of this equation leads to:

$$\varepsilon_r = \frac{1}{{k_0}^2}({k'_q}^2 + j2k'_q k''_q - {k''_q}^2 - \alpha^2 - j2\alpha\beta + \beta^2) \tag{2.52}$$

with $\varepsilon_r = \varepsilon'_r - j\varepsilon''_r$.

The dielectric constant and the loss factor can be calculated by:

$$\varepsilon'_r = \frac{1}{{k_0}^2}({k'_q}^2 - {k''_q}^2 - \alpha^2 + \beta^2) \tag{2.53}$$

$$\varepsilon''_r = \frac{1}{{k_0}^2}(2\alpha\beta - 2k'_q k''_q) \tag{2.54}$$

The formulas 2.53 and 2.54 are the same as the corresponding formulas in [AFT07]. In the present work, the calculation of dielectric properties with these formulas is compared with the neural network characterization and presented in Chapter 4.

3 Method and simulation model

This Chapter describes investigations on an appropriate model for the measurement setup. It starts from the simulation of the measurement setup to obtain optimal crucible size. Then dielectric reconstruction using a neural network is investigated.

3.1 Simulation

Full 3D electromagnetic wave simulations using CST Microwave Studio (MWS) [Cst] are performed to optimize the measurement setup for temperature and time dependent dielectric measurements on thermosetting resins. The dataset of the proposed neural network optimization is also obtained from the selected 3D model of the measurement setup. CST MWS is a commercial finite integration time domain code.

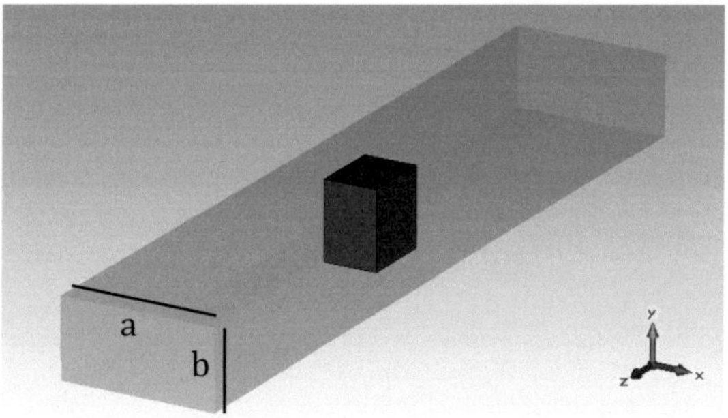

Figure 3.1: Schematic of measurement setup

Fig. 3.1 shows the 3D model of the measurement setup. The sample crucible is modelled with adjustable length and width to accommodate the simulation of different sizes. Teflon is selected as the resin container. Teflon has low dielectric constant and loss factor so that it does not reduce the sensitivity of the measurement. It is also easy machined and can be used up to 150°C. The dimension of the crucible is optimized to achieve good sensitivity. Considering experimental aspects, 5 mm thickness of the container is chosen. The thinner the wall is, the higher is the possibility for deformations of the crucible; on the other hand, the thicker it is, it may reduce the sensitivity. In order to acquire a sensitive measurement system, different sizes of the resin crucible are simulated.

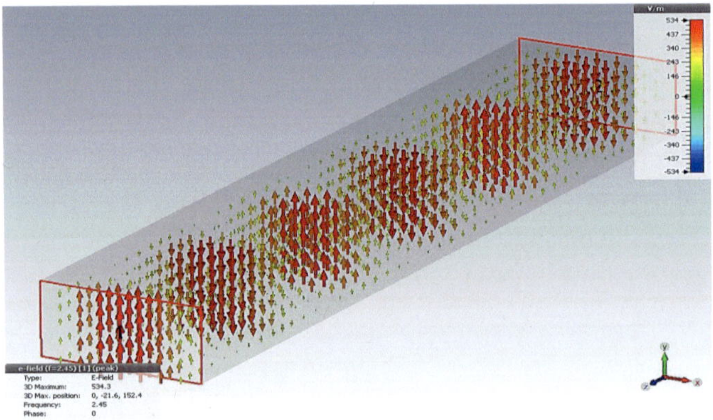

Figure 3.2: Electric field of the measurement waveguide without sample

In Fig. 3.2 the electromagnetic wave propagates from port 1 to port 2. The magnitude of the electric field is represented by different colors. The strongest electric field is in the middle of the broad side a of the waveguide cross section in x direction. This fact means that the sample should be placed in the middle of the broad side. Introducing a sample (Teflon) in the waveguide between ports 1 and 2 causes changing of the electric field after the sample position, which can be seen in Fig. 3.3.

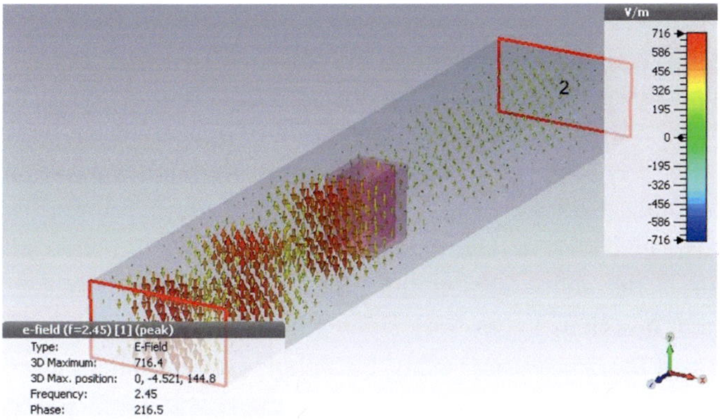

Figure 3.3: Electric field of the measurement waveguide with inserted Teflon (20 mm x 43.2 mm x 30 mm)

The change of the electric field depends on the dielectric properties of material that is inserted in the system. Of course the changing of the electric field will have an influence on the electromagnetic wave at the ports of measurement. The transmission and reflection coefficients are changed, depending on the changing of electric and magnetic fields. The dielectric properties of the sample can be calculated by reconstruction using the measured *S*-parameters. Since only nonmagnetic samples are investigated, changing of the magnetic field is not considered here.

The size of the sample influences the sensitivity of the measurement. In fully covered waveguide, the thickness of the sample should be matched with the material properties to achive better sensitivity. Low loss material needs thicker samples, while high loss material needs thinner samples. Like in a fully coverered waveguide, in the partially covered waveguide, the optimal size of the sample for dielectric measurements depends on the dielectric properties of the sample.

In this study, the crucible size is decided based on practical consideration such as the required volume and a uniform heat distribution of the sample. Considering temperature dependent measurements, the size of the sample should be as small as possible. Small size is required in order to achieve fast uniform heating of the sample. However, the sensitivity of the measurement should be also considered. The smaller the size of the crucible is, the lower is the sensitivity of the measurement, especially for loss tangent less than 1 [Con+04]. Since the dielectric constant and loss factor of thermosetting resins, hardeners and bio fillers are commonly ranging from $\varepsilon_r' = 2\text{-}10$ and $\varepsilon_r'' = 0.01 - 2$, respectively, the crucible is optimized for this parameter range of dielectrics. In Fig. 3.4 a small crucible with the size of 10 mm x 43.2 mm x 20 mm (inner dimensions) is introduced in the waveguide sample holder.

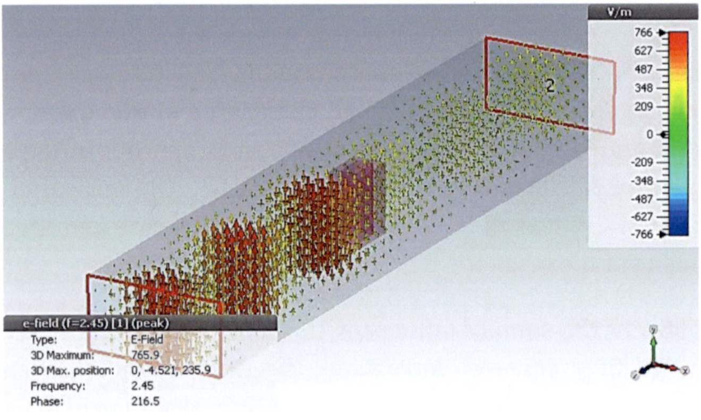

Figure 3.4: Electric field of the measurement waveguide with sample (10 mm x 43.2 mm x 10 mm)

By comparing Fig. 3.3 and 3.4, a larger change of the electric field is found for the larger sample. The electric field at Port 2 becomes much weaker in the larger sample compared to the smaller sample. In the experiment, scattering parameters are measured instead of electric field strength. Different size of sample holder is simulated to investigate *S*-parameter differences. The transmission and reflection coefficients for various sizes of the filled sample holder are compared with the empty Teflon crucible as a reference. Smaller difference causes less sensitivity. In the neural network, *S*-parameters are considered as scalars. Thus, both differences in the real and imaginary part of the *S*-parameters are considered as scalars.

The S_{21} differences depend on the size of the crucible as shown in Table 3.1. The larger the size of the crucible, the larger is the difference of S_{21}. Magnitude and phase of the *S*-parameters show the same behavior.

Table 3.1: Differences of S_{21}-parameter between the empty crucible (reference) and the filled one with the sample characterized by $\varepsilon_r' = 3.8$ and $\varepsilon_r'' = 1.2$

Inner dimension of crucible (x.y.z mm³)	Real part difference	Imag. part difference	Magnitude difference	Phase difference (°)
30x43.2x30	0.1577	0.8344	0.3812	62.66
20x43.2x30	0.0710	0.7215	0.3386	47.50
20x43.2x20	0.0454	0.4924	0.3174	27.54
10x43.2x20	0.0334	0.3402	0.2596	17.41
10x43.2x10	0.0216	0.1901	0.1406	9.95

The features of the differences in the S_{11}-parameters for the different crucibles cannot be concluded from the reflection coefficient (see Table 3.2). For samples with relatively high loss factor, the 20 mm x 43.2 mm x 30 mm crucible makes a significant difference to an empty crucible both in transmission and reflection compared to other crucibles.

Table 3.2: Differences of S_{11}-parameter between the empty crucible (reference) and the filled one with the sample characterized by $\varepsilon_r' = 3.8$ and $\varepsilon_r'' = 1.2$

Inner dimension of crucible (x.y.z mm³)	Real part difference	Imag. part difference	Magnitude difference	Phase difference (°)
30x43.2x30	0.0553	0.1750	0.1417	34.79
20x43.2x30	0.1005	0.2391	0.1779	41.31
20x43.2x20	0.1075	0.3753	0.3073	40.34
10x43.2x20	0.0315	0.2878	0.2329	31.39
10x43.2x10	0.0182	0.2076	0.1728	23.24

Simulations for moderate dielectric loss factor (1.2) are performed in order to see the *S*-parameter behavior of thermosetting resins. During the heating from room temperature to the curing temperature, thermosetting resins commonly have moderate dielectric loss factors. The simulation shows that samples in larger crucibles lead to larger differences of the *S*-parameters. This fact means that in order to better investigate the changing of loss factors during the heating period, larger crucibles are preferable.

Further simulations with relatively low loss factors are required to investigate cured polymer composites which commonly also have relatively low loss factor. Simulation of a material with $\varepsilon_r'' = 0.02$ was performed and is shown in Table 3.3.

Table 3.3: Differences of S_{21}-parameter between the empty crucible (reference) and the filled one with the sample characterized by $\varepsilon_r' = 3.8$ and $\varepsilon_r'' = 0.02$

Inner dimension of crucible (x.y.z mm³)	Real part difference	Imag. part difference	Magnitude difference	Phase difference (°)
30x43.2x30	0.2702	0.9675	0.0413	61.08
20x43.2x30	0.1993	0.7331	0.0780	47.07
20x43.2x20	0.0984	0.5016	0.1829	31.67
10x43.2x20	0.1202	0.3259	0.1124	20.72
10x43.2x10	0.1014	0.1897	0.0689	12.47

The results show that larger crucibles have a larger difference both in the real and imaginary part of the S-parameters. The same behavior can also be found in the phase difference. However, the difference in magnitude does not follow this feature. The same features are also found for S_{11} as shown in Table 3.4.

The simulations show that the difference of the transmission coefficients is consistent with the crucible volume both for the real and imaginary part of the S-parameters. Larger sample size makes larger difference. In the reflection coefficient, both real-and imaginary part as well as magnitude and phase show an inconsistent behavior. The changing of S-parameters in the smaller size crucible tends to be less compared with larger ones for a chosen value range of dielectric properties. However, it cannot be generally concluded that a larger size of the sample has better sensitivity, because this depends on the range of the dielectric properties to be measured.

Table 3.4: Differences of S_{11}-parameter between the empty crucible (reference) and the filled one with the sample characterized by $\varepsilon_r' = 3.8$ and $\varepsilon_r'' = 0.02$

Inner dimension of crucible (x.y.z mm³)	Real part difference	Imag. part difference	Magnitude difference	Phase difference (°)
30x43.2x30	0.1783	0.3786	0.1541	59.62
20x43.2x30	0.1342	0.2863	0.2428	46.37
20x43.2x20	0.0059	0.4310	0.381	31.46
10x43.2x20	0.0696	0.2963	0.28	20.589
10x43.2x10	0.0960	0.1856	0.1989	112.32

In the following studies, the sample size of 20 mm x 43.2 mm x 30 mm is selected. The larger one is not preferred because it needs more volume of MUT. The sample size 30 mm x 43.2 mm x 30 also does not satisfy the criteria of thin sample $d \leq 0.25a$ [AFT07, CCFD03, Has06]. The larger sample also needs longer time to reach uniform thermal distribution during heating. Fast heating rates are not preferable due to the risk of strong exothermic reaction. Fast heating also causes trapping of the molecules of resins and hardeners so that the fully cured status cannot be reached [NND04].

3.2 Neural network

In [Hay98], a neural network is defined as an algorithm that operates in the same way as a biological brain solves problems. So a neural network is an algorithm inspired by the ability of the human brain to learn many tasks. A neural network consists of units, nodes and neurons that communicate to each other. Neural networks find broad applications such as in control systems, computer visions and other fields. In literature, neural networks are also applied for solving electromagnetic problems [YME05, OPRT02 and PFSJ03]. In order to implement a neural network for characterization of dielectric properties, a structure and learning algorithm has to be defined. Input and output of a neural network depends on the real system. Many experimental aspects have to be defined before the implementation of a neural network.

The goal of this optimization using a neural network is to provide a fast calculation of the dielectric properties. In temperature and time dependent measurements, the requirement of a fast algorithm for calculation of dielectric properties is compulsory. First, the network is trained appropriately within a certain range of dielectric properties. Considering the measured materials, the ranges of 1.0-10 and 0.01-10 are used for the real and imaginary parts of the complex relative permittivity, respectively.

The second step is testing with a new dataset and finally with experimental data. The network needs sufficient knowledge in order to reconstruct the dielectric properties of a material. The dataset for training and testing is generated specifically to measure the dielectric properties of thermosetting resins during curing. A well-defined range of dielectric properties is required for generating a database according to any specific process. The well-defined database makes the network easier to reconstruct the dielectric data. The database generator is implemented by running the CST MWS model with a Matlab script for the selected range of dielectric properties with uniform random distribution. Uniform random distribution of dielectric properties is required to achieve generalized relations of scattering parameters and dielectric properties.

The training inputs for the neural network are the simulated S-parameters. Therefore the experimental setup is modeled using CST MWS so that the scattering parameters for different values of dielectric constant and dielectric loss factor can be obtained. The input can be a pair of $|S_{11}|$ and $|S_{21}|$ parameters or all the real and imaginary parts of S_{11} and S_{21}. In [YME05], the real and imaginary values of S_{11} and S_{21} are selected as input. The sample is located in the narrow section of the rectangular waveguide. A different scheme of input can be found in [PFSJ03] where only the magnitudes of S_{11} and S_{21} were selected for the input of dielectric properties reconstruction.

The input parameters depend on the method of reconstruction of the dielectric properties. The reflection method only needs the S_{11}-parameters for the input. In the present study, the transmission and reflection method is selected so that the full two port data will be used for the input. Omitting some parameters or transforming to other forms might be required to increase the convergency of the network or the accuracy of the reconstruction. In this work, based on the simulation results, the real and imaginary parts of S_{11} and S_{21} are used.

3.2.1 Structure of the network

In [CP43] the structure of a simple logical unit has been presented, that is now considered as early development of neural networks. A further study of McCulloch-Pits neurons yields the threshold logic unit with adjustable weight [Hay98]. In Fig. 3.5 a simple neural network with n inputs and one output is shown.

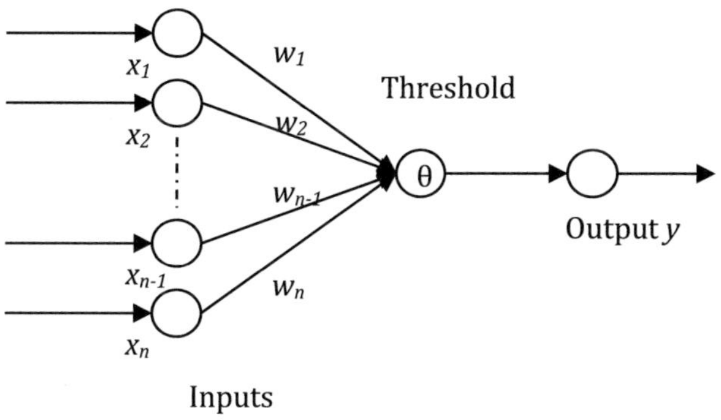

Figure 3.5: Simple neural network

In the above simple network, the input, weight and output parameters are the variable *x, w* and *y*, respectively. The structure gives (0 or 1) output based on the threshold of the sum of inputs multiplied by the weights.

$$y(p+1) = \begin{cases} 1, & if \sum_{i=1}^{n} w_i x_i(p) \geq \theta \\ 0, & otherwise \end{cases} \tag{3.1}$$

In recent neural networks, the threshold function is popularly known as activation function. Many functions are in use for the activation function in neural networks:

- Hard limiter function
  The hard limiter or threshold function can be written as

$$f(x) = \begin{cases} 1, & \mathrm{x} \geq x_1 \\ 0, & otherwise \end{cases} \tag{3.2}$$

A constant x_1 is a threshold value.

- linear function

$$f(x) = x \tag{3.3}$$

- Sigmoid function

$$f(x) = \frac{1}{1+e^{-x}} \tag{3.4}$$

- Hyperbolic tangent function

$$f(x) = \frac{1-e^{-x}}{1+e^{-x}} \tag{3.5}$$

A multilayer perceptron (MLP) and a radial basis function (RBF) neural network are often selected in approximation, prediction and pattern recognition problems [Hay98]. In [EMY07] a RBF was chosen as a neural network structure to calculate dielectric properties of liquids. Dielectric properties of liquids at room temperature are calculated using this neural network model, which is trained by S-parameters simulated by the FDTD method. The measurement setup was not implemented in the measurement of temperature dependence.

In [PFSJ03 and OPRT02], a MLP was selected for dielectric properties reconstruction. Both RBF and MLP neural networks are able to reconstruct dielectric properties of the material. The MLP and RBF neural networks are compared in this present work for reconstruction of dielectric properties as described at the end of this subchapter.

In the RBF model (see Fig. 3.6), the input layer and the hidden layer are connected directly without weighting factor [YME05]. Instead, a center vector (U) is used in the hidden layer. If p refers to the pth hidden neuron, the output of the hidden layer (h_p) with a vector input S is:

$$h_p\,(n) = f_p(\| S(n) - U_p \|) \text{ with } \| \cdot \| = \text{Euclidean norm} \tag{3.6}$$

where n is a time sequence and f_p is the function of the hidden neuron p. The transfer functions of the hidden neurons can be a [Hay98]:

- Gaussian function: $f_p(L) = e^{-\frac{L^2}{\sigma^2}}$,
- multiquadratic function: $f_p(L) = (L^2 + \sigma^2)^{\frac{1}{2}}$,
- piece-wise linear function: $f_p(L) = \frac{1}{2}(|L+1| - |L-1|)$, or a
- cubic approximation function: $f_p(L) = \frac{1}{2}(|L^3+1| - |L^3-1|)$.

The variable L is the distance between the input vector and the center vector, and σ is a scaling parameter.

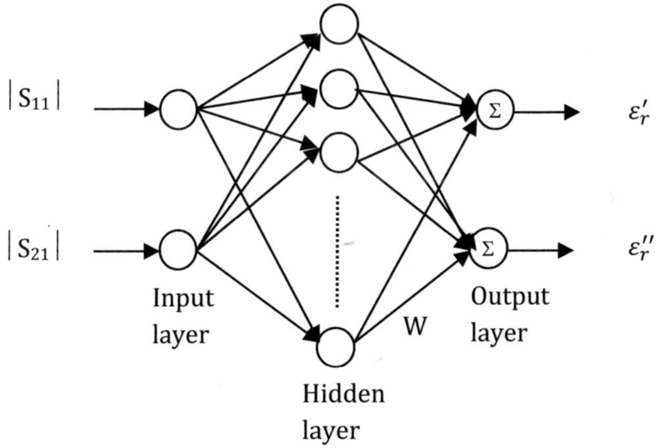

Figure 3.6: Scheme of a radial basis function (RBF) neural network for dielectric reconstruction

The output of the vth neuron in the layer output ε_r^v can be de- described by:

$$\varepsilon_r^v(n) = \sum_{p=1}^{P} w_{pv} h_p(n) = \sum_{p=1}^{P} w_{pv} f_p(\| S(n) - U_p \|) \quad (3.7)$$

where P is the number of hidden neurons, and $v = 1,2,\dots V$ (V is the number of the output neurons). An MLP network with two hidden layers is shown in Fig. 3.7.

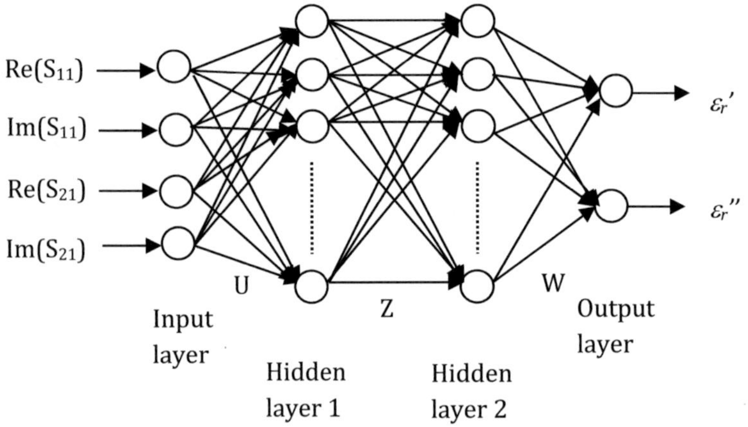

Figure 3.7: A multi-layers perceptron (MLP) neural network for reconstruction of dielectric properties

The MLP network output can be calculated as:

$$\varepsilon_r^v = g\left(\sum_{p_2=1}^{P_2} w_{p_2 v} f_2\left(\sum_{p_1=1}^{P_1} z_{p_1 p_2} f_1\left(\sum_{i=1}^{4} u_{i p_1} S_i\right)\right)\right) \quad (3.8)$$

where:

S_i and ε_r^v are the input and output vectors of the network,
g is the activation function of the output neurons,
f_1 and f_2 are the activation functions of the hidden neurons in layer 1 and 2,

U, Z, W are the weight vectors, and
P_1 and P_2 are the number of the hidden neurons in layer 1 and layer 2.

Neural networks have been very often used as tools to solve electromagnetic problems. Many publications also reveal similar findings in this field [OPRT02, PFSJ02 and YMW05].

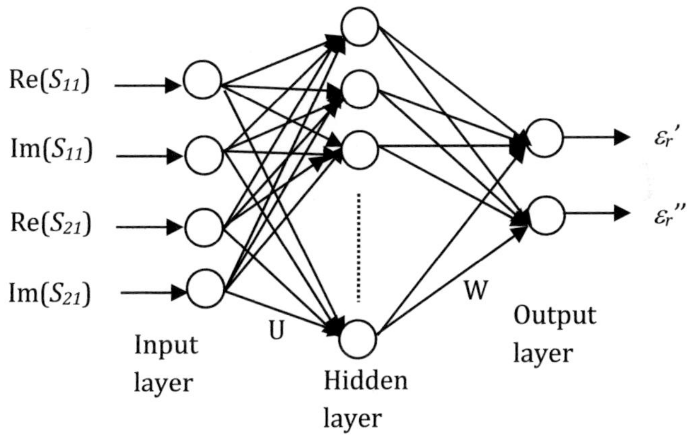

Figure 3.8: Neural network model for reconstruction of dielectric properties in [YME05]

Reconstruction of the complex permittivity is reported in [EMY07] by using FDTD modeling and neural networks. The network is trained by using permittivity data and scattering parameters of the material simulated using the FDTD method. A formula can be derived from the structure:

$$\varepsilon_l = g\left(\sum_{j=1}^{N} w_{jl} h(\sum_{i=1}^{4} S_i)\right) \tag{3.9}$$

where S is the input vector, the function h is the activation function of layer 1, w is weight between the hidden layer and the output layer and g is the activation function of the output layer.

The input vector is given by:

$$S = (S_1 \quad S_2 \quad S_3 \quad S_4) = (Re(S_{11}) \quad Im(S_{11}) \quad Re(S_{21}) \quad Im(S_{21})) \tag{3.10}$$

In the present work, the neural network is applied to reconstruct the dielectric properties of materials during curing. As already mentioned, in order to select an appropriate structure, the RBF and MLP neural network are compared. The CST MWS code gives a database needed for the reconstruction of dielectric properties by a neural network. Then the neural network algorithm performs the reconstruction of the dielectric properties with simulated data as the training dataset.

The multilayer perceptron with the back propagation error method is selected to solve inverse S-parameter problems. The multilayer perceptron network consists of input layer, output layer and hidden layers. The different configurations of hidden layers are always testable to get the best result of characterization. A large number of hidden layers of course influences the time needed in training.

3.2.2 Network training and testing

A neural network has to be trained before it can be used to perform some tasks. The training methods can be classified into:

- Supervised learning
  The training is performed using input-output pairs provided by an external teacher or by the system itself. The actual output for each input pattern guides the training process. Weights of the neural network layers are adapted during training to match the desired output.

- Unsupervised learning
  The output units are adapted to the cluster of the input patterns. The system should discover statistically the patterns of the input.

In the characterization of dielectric properties, the supervised learning is suitable. The external teacher is represented by the 3D CST MWS electromagnetic wave simulations of the measurement setup as the source of the training database generator. The training can be performed online or offline. However, in order to avoid unnecessary long required training time during the real monitoring of curing, the offline supervised training is preferable. The neural network is trained by using sufficient training data sets before it is used to calculate dielectric properties of materials.

The structure of MLP and RBF neural networks are optimized to get the minimal mean square error of target and output values. An iteration procedure is written in Matlab script [Mat] to obtain the best structure that represents the system. The number of hidden layers is limited to three layers for reducing the optimization time. The developed measurement setup is designed as simple as possible so that the neural network can easier recognize the relation between dielectric properties and S-parameters.

In order to achieve optimal neural network training, the dataset is divided into three subsets. The first one is the subset for computing the gradient and updating the weight biases. This subset is mainly used to train the network so that the relationship between S-parameters and the dielectric properties can be represented with the structure and weight biases of the network. The second subset is the validation set for monitoring any errors during training. This validation set prevents over-fitting of the data. The last subset is the test set to compare with other structures. The selection of dielectric properties in the defined range has to have a uniform random distribution.

This uniform random distribution is the representative of the population of dielectric properties within any defined range.

An example of training performance of a MLP network is presented in Fig. 3.9.

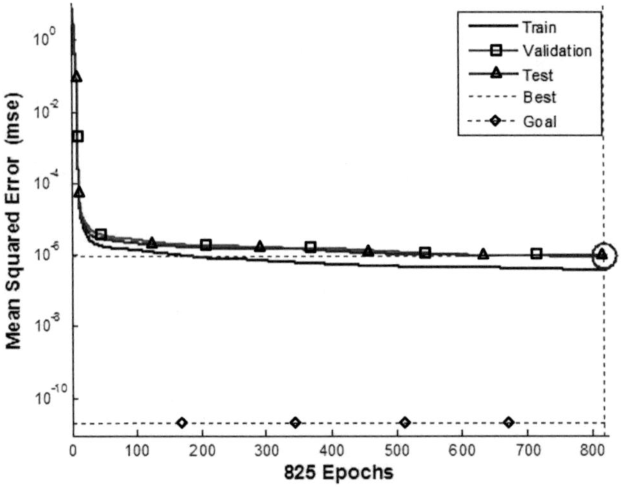

Figure 3.9: Training performance of a MLP neural network

The neural network is trained, validated and tested with 1000 pairs of dielectric properties and scattering parameters (700, 150 and 150 pairs for training, validation and testing respectively). In this training, the best training performance gives a mean squared error (MSE) of 9.3E-7 at epoch 819. The MSE of training at epoch 825 is 6.2E-7. The test and validation MSE asymptotically goes to approximately 1E-6. The evaluation with test data is performed and shown in Fig. 3.10. and Fig. 3.11. The network is tested with 150 pairs of S_{11} and S_{21} which are not used in the training.

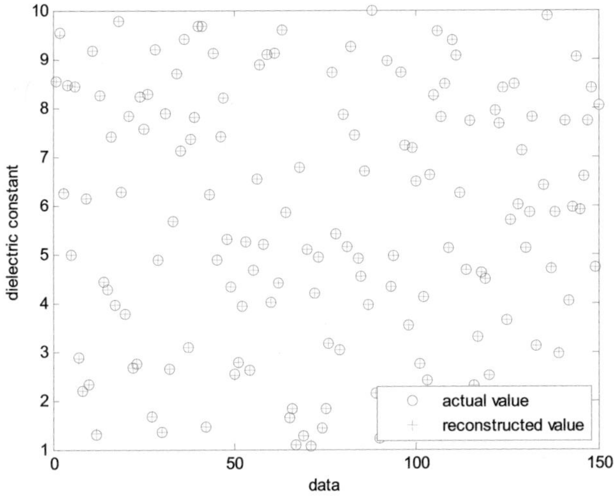

Figure 3.10: Testing of dielectric constant reconstruction of MLP neural network.

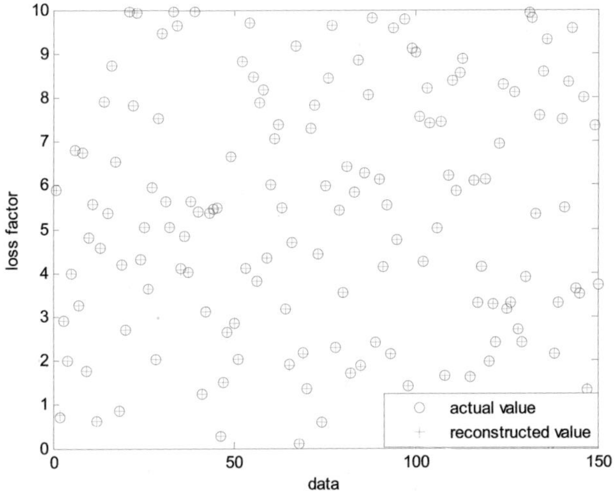

Figure 3.11: Testing of loss factor reconstruction of MLP neural network.

The target (dielectric constant and loss factor) is also in the range of $\varepsilon_r' = 1$ to 10 and $\varepsilon_r'' = 0.01$ to 10, respectively with random uniform distribution. The output of the network fits well with the target which indicates that the network is able to characterize the *S*-parameter input to get the dielectric properties.

In order to make a clear difference, testing with few dielectric property values is performed and presented in Table 3.5.

Table 3.5: Neural network testing with simulated *S*-parameters

Actual Value		Reconstruction using MLP		Reconstruction using RBF	
Dielectric constant	Loss Factor	Dielectric constant	Loss Factor	Dielectric constant	Loss Factor
2.03	0.01	2.02	0.01	2.02	0.02
3.79	0.70	3.80	0.70	3.82	0.72
7.80	1.20	7.80	1.20	7.83	1.19

Both the RBF and MLP networks are tested with the same dielectric properties. The result shows that both MLP and RBF are able to calculate the dielectric properties from testing datasets. MLP has a slightly better performance as compared to RBF. In order to apply the neural network reconstruction to the experimental data, a validation with well-defined materials is required. The validation is also performed by measurement of materials with other methods. Here the measurement with a coaxial probe is selected to validate the measurement result. Detail validation with experimental data is explained in Chapter 4.

4 Description of the measurement setup and investigated polymer composites

In this Chapter the experimental setup and the investigated polymer composites are described. In the measurement setup the measurement system is explained followed by the automatic control system. Then, the validation of the measurement setup and estimation of error are presented. Finally, the selection of the materials for experimental investigations is discussed.

4.1 Measurement setup

The automatic system for curing monitoring with 2.45 GHz dielectric analysis (DEA) has been designed as shown in Fig. 4.1:

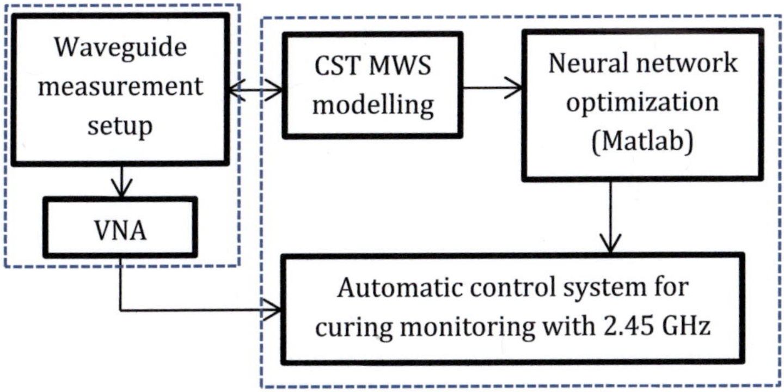

Figure 4.1: Block diagram of the dielectric monitoring system for curing of polymer composites

Realization of the proposed measurement setup is started by modelling the waveguide measurement setup using CST MWS. Reciprocal steps of simulation and measurement have been done for obtaining the optimal setup. The second step is optimization of a neural network to represent the modelled measurement setup. The neural network has been trained and tested with the model of the measurement setup as explained in Chapter 3. The third step is validation of the measurement setup. The neural network also has been validated with experimental data which is described in subchapter 4.2. The last step is design and implementation of the automatic system for curing monitoring. The system collects the *S*-parameter data of the measured sample from the VNA, and calculates the dielectric properties during curing.

4.1.1 Measurement system

The measurement setup consists of the waveguide sample holder, additional waveguide sections, waveguide-to-coaxial adapters, coaxial cables, Agilent PNA 5224 vector network analyzer (VNA) and personal computer (PC). The additional 200 mm long waveguide sections are installed on both ports connected to the waveguide-to-coaxial adapter in order to reduce the effect of higher order (evanescent) modes in the measurement result. Their length is approximately 1.5 times the free-space wavelength at 2.45 GHz. The higher order evanescent modes vanish at the end of the additional waveguide. As an example the first higher order mode TE_{20} is strongly evanescent with an attenuation factor of 448 dB/m at 2.45 GHz. Then with 200 mm additional waveguide, the higher order evanescent mode is attenuated by 89.6 dB. The setup is also equipped with a pneumatic system that allows to connect and disconnect the the additional waveguides to the waveguide sample holder ports by automatic control. The waveguides are connected, that means closed, during measurements and disconnected, that means opened, at any other time. This opening-closure system prevents significant heat transfer from the hot waveguide sample holder to the measurement set-up that may influence the system calibration. A thermocouple type K is installed in the middle of the waveguide sample holder through a small hole with 0.5 mm radius to measure the temperature at the materials surface. Another thermocouple is installed in the bottom of the waveguide to check the correct temperature. Four resistive heating elements with total power 300 W are installed in the waveguide sample holder. The measurement setup is shown in Fig. 4.2. The VNA is connected through coaxial cables with the waveguide to coaxial adapters.

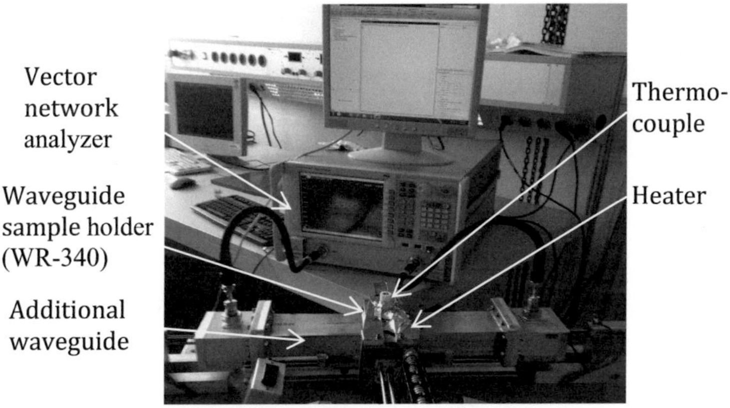

Figure 4.2: Measurement setup

4.1.2 Control system for automatic measurements

The measurement system is designed to measure the dielectric properties of the material during curing. The development of a control system for heating and data acquisition is significant to obtain reliable data. Measurements on temperature and time dependent dielectric properties would be time consuming and very difficult without an automatic system. In order to design a fully automatic monitoring system for the curing process, the following requirements are needed:

- Automatic control of the material temperature based on a preset temperature schedule.
- Automatic data acquisition of the material temperature.
- Automatic data acquisition of the *S*-parameters.
- Automatic calculation of the dielectric properties.
- Automatic data acquisition of the curing time.

These requirements are fulfilled using a computer-based automatic control and data acquisition system. A simple block diagram of the system based on the above requirements is shown in Fig. 4.3:

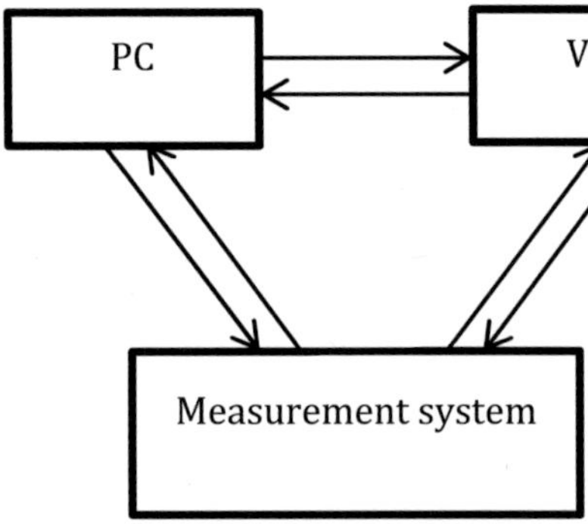

Figure 4.3: Block diagram of the present automatic measurement system

The communication between PC and measurement system is bidirectional, recurring of:

- Acquisition of temperature data.
- Acquisition of waveguide position.
- Sending command for position and temperature control.

The communication between PC and VNA is also bidirectional, i.e. sending command and data acquisition for *S*-parameter measurement. The relation between VNA and the measurement setup is the measurement of the *S*-parameters through the waveguide-to-coaxial cable adapters. As already mentioned, the Agilent VNA PNA 5224 is selected for the measurements. The PNA 5224 has a well-defined accuracy for *S*-Parameter measurements with noise floor down to -120 dB. Through reflect line (TRL) calibration which is needed in waveguide measurements is also available in this VNA. The measurement system has a dynamic range of 80 dB using an IF bandwidth of 1 kHz.

The proposed automatic system can be separated into two parts. The first one is the system for position control and acquisition of the *S*-parameter data. The second is the system that controls and acquires temperature data of the material. The block diagrams of these two systems are shown in Fig. 4.4 and Fig. 4.5, respectively.

The position control is a feed forward control system. The feed forward system works based on predefined way to respond the control signal without any consideration of how the load reacts to the signal. The position sensor uses only two conditions (on and off). The PC collects the condition of the sensors to set the initial and measurement position. The information of the position is transferred to the PC by use the digital input and output channels of a National Instrument data acquisition card type PCI 6220. This information is used to trigger the *S*-parameter measurements. An *S*-parameter measurement is only performed when the waveguide port already has a contact with the waveguide sample holder.

The temperature acquisition and control system is shown in Fig. 4.5. The difference between set value and measured temperature is forwarded to the Propotional Differential and Integral (PID) controller. The output of the PID controller is used as the input for the heating system. Therefore the PID output signal is converted to a pulsed signal with modulated pulse width.

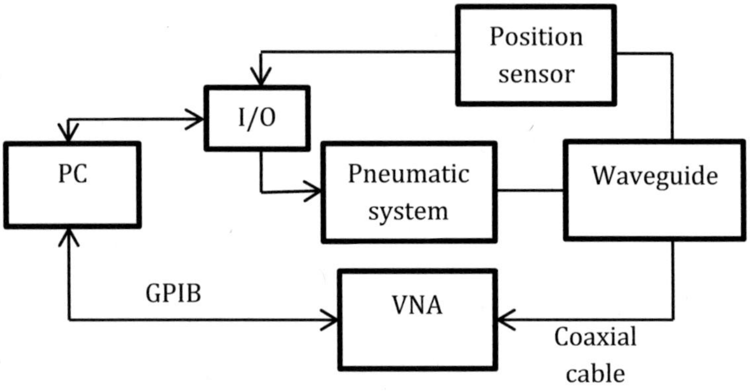

Figure 4.4: Block diagram of position control and *S*-Parameter data acquisition

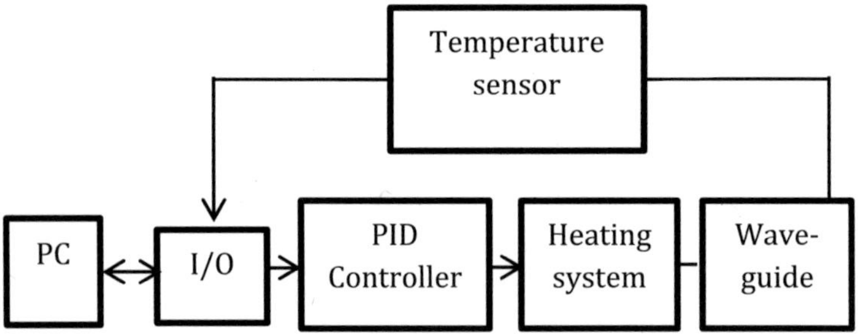

Figure 4.5: Block diagram of temperature control and acquisition

The temperature of the sample is measured on its surface. In order to have a uniform distribution of the heating, a slow heating rate (maximum 1°C per minutes) is applied. The measured temperature is compared with user-defined set values.

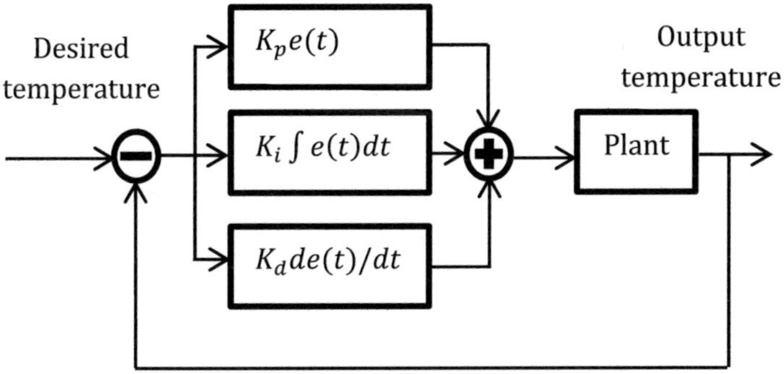

Figure 4.6: Block diagram of PID closed loop control system

The output of the PID controller $u(t)$ in Fig. 4.6 can be written as [AM08]:

$$u(t) = K_p e(t) + K_i \int e(t)dt + K_d e(t)/dt \qquad (4.1)$$

where $e(t)$ is the error signal. The proportional, integral and derivative gains are K_p, K_i, and K_d, respectively.

In the proposed system, the set values are derived from the temperature profile with subsequent regimes of constant heating rates and isothermal curing which has to be defined before the measurement of curing starts. The sample temperature is continuously measured by a thermocouple. The difference between output and desired temperature is used for the calculation of the PID controller output that drives the actuators which are the 4 resistive heating elements.

The parameter of the PID controller can also be altered during the heating process to allow fast optimization of the PID parameters. The optimal PID parameters obtained for the control system are K_p = 0.9, K_d = 0 and K_i = 0.003. These parameters are obtained using online tuning of the parameters during heating. By use of these parameters an excellent temperature control can be demonstrated with maximum temperature deviation of less than 0.2°C as shown in Fig. 4.7.

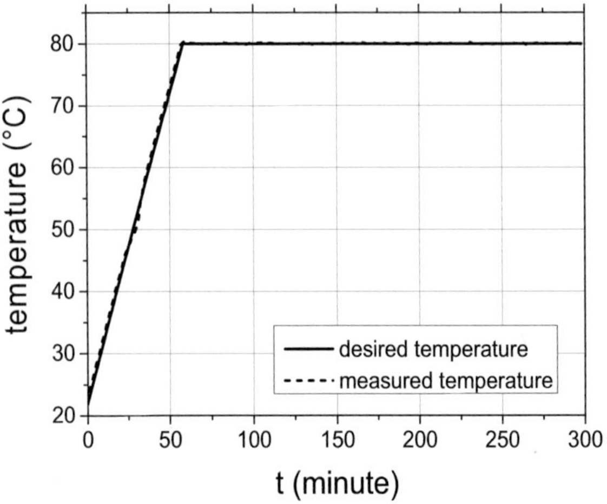

Figure 4.7: Temperature controlled by optimized PID system

In order to make an easy handable measurement system, a graphical user interface (GUI) for setting the measurement procedure has been developed which is shown in Fig. 4.8. In the GUI, the scheme of curing monitoring can be freely set. The system allows flexible schemes which include the characterization of curing as close as possible to the real curing. The flexible scheme also allows experiments for specific purpose such systematic investigations of the influence of temperature, hardeners and reinforcement. The temperature can be set, either as stepwise temperature scheme or constant temperature scheme. The system also allows a linear ramp scheme (minimal 0.3°C/min) from room temperature to a specific constant temperature. The rate can be flexibly changed according to the characteristic of the hardeners.

Commonly a slow heating rate is set during curing because the properties of the hardeners (curing agents) usually do not allow fast heating.

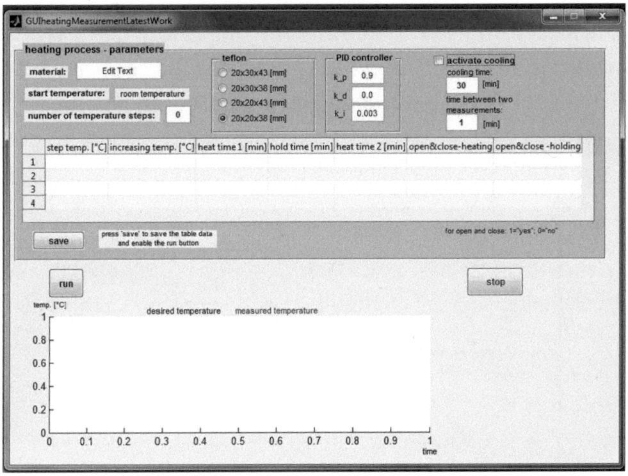

Figure 4.8: Graphical User Interface (GUI) of the measurement setting

Slow heating rates can avoid excessive exothermal reaction. In addition, the maximal rate is also limited in practice because the heat transfer requires time to reach a uniform temperature distribution. In Fig. 4.9, typical temperature schemes for curing are presented. The first scheme is a constant temperature scheme that needs a preheating period in the real curing. The second scheme is a ramp-constant temperature scheme. This scheme can be implemented in real curing since the heating period to constant temperature is included. The last one, the step-wise scheme, is the common scheme in conventional curing to ensure the full curing of resins with conventional heating. The designed GUI also allows the user to select a Teflon crucible. The system has already been optimized for the measurement of real and imaginary part of the relative complex permittivity in the range of 1.0-10 and 0.01-10, respectively. The selected range of the dielectric properties is commonly true for thermosetting resins and hardeners. The Teflon holder is optimized at the inner dimension size of 20 mm x 43.10 mm x 30mm. However, the flexibility is still offered in the GUI by another size of the Teflon crucible that permits the user to measure relatively high dielectric loss factor by choosing a smaller Teflon crucible. The opening- closure system can also be activated or deactivated in the GUI.

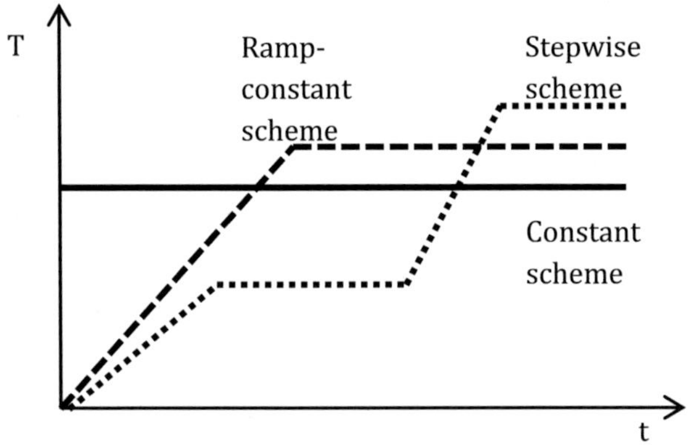

Figure 4.9: Different temperature schemes for curing

The system is also designed for measurements during the cooling phase that can be flexibly set with respect to period and duration of the measurement. All the flexible parameters benefit the investigation of curing according to its special condition. The procedure of a curing process measurement is described in the following steps:

- Doing the Through Reflect Line (TRL) calibration
- Inserting a Teflon crucible filled with the mixture of materials into the waveguide sample holder
- Setting the temperature and time scheme
- Starting the automatic heat control and curing monitoring

4.2 Validation of the measurement setup

The measurement setup was validated to achieve accurate measurement results. The validation of the neural network algorithm based on simulated data has been explained in Chapter 3. In this subchapter, the validation is conducted based on experimental data. The proposed method of reconstruction of dielectric properties is based on the modeling of the measurement setup using CST MWS. This model is utilized for generating datasets for dielectric parameter reconstruction employing the neural network algorithm.

The validation of simulated and measured *S*-parameters of some materials has been carried out. Here the example of the validation with an empty Teflon crucible with inner dimension 20 mm x 43.20 mm x 30 mm is shown in Fig. 4.10 and 4.11.

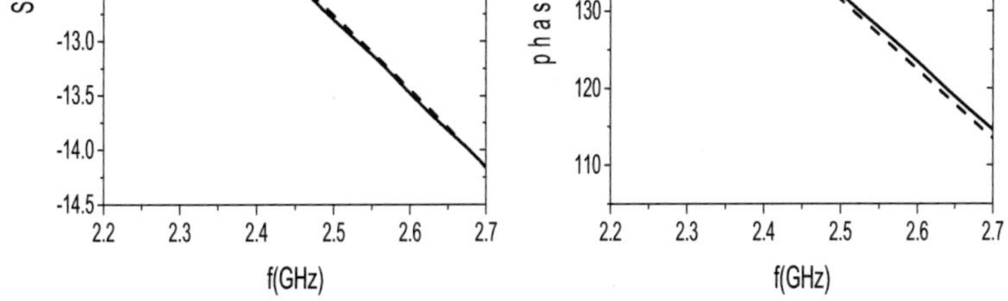

Figure 4.10: Simulated and measured magnitude and phase of S_{11} of an empty Teflon crucible

The simulated and measured S_{11} of the measurement setup with an empty Teflon crucible are in very good agreement in both magnitude and phase, as shown in Fig. 4.10. The S_{21}-parameter also fits well in both magnitude and phase, as shown in Fig. 4.11.

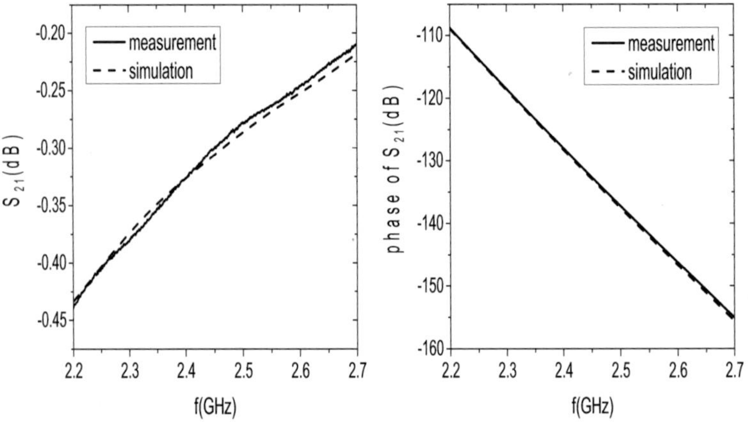

Figure 4.11: Simulated and measured magnitude and phase of S_{21} of an empty Teflon crucible

The dielectric measurements using the proposed measurement setup are validated by measurements on well-known materials and on new materials considered for the present experiments at room temperature. All materials were also measured by using a coaxial probe HP85070B [HP97] as reference method for lossy materials. The validation results are presented in Table 4.1.

In the measurements on resin and hardeners, the proposed method shows good agreement with coaxial probe results for both dielectric constant and loss factor. The coaxial probe can not measure low loss material as can be seen from Table 4.1. The analytical solution [AFT07] based on formulas 2.50, 2.53 and 2.54

given in Chapter 2 has also been tested. The results show that the analytical method is suitable for low loss materials.

Table 4.1: Validation with well-known and new materials

Material	Proposed method		Coaxial probe		Analytical method	
	ε_r'	ε_r''	ε_r'	ε_r''	ε_r'	ε_r''
PTFE	2.0	0.01	1.9	NA	2.2	0.01
PVC	2.8	0.02	2.7	NA	2.8	0.02
PMMA	2.6	0.03	2.5	NA	2.6	0.02
DGEBA	3.7	0.68	3.8	0.70	3.5	0.52
PEA	4.5	1.40	4.5	1.45	4.2	0.93
IPDA	4.3	0.85	4.3	0.87	4.0	0.64
Laromin	4.4	1.12	4.4	1.14	4.1	0.88

In the measurement of lossy material such as resin and hardeners, the analytical method has larger errors compared to the proposed method and the coaxial probe result.

In order to clearly show the accuracy of the proposed method compared to the analytical method, validations on specific range of dielectric properties have been done. Since the actual values of dielectric properties of materials are only based on the reference method in the validation with the measured data, the simulated data is selected for further validation. The dielectric properties are given as inputs in the simulation model to generate *S*-parameter data. The given dielectric properties in the simulation model are considered as actual values. Then, the corresponding *S*-parameters are considered as inputs for reconstruction dielectric properties using the neural network (proposed method) and the analytical method. The validation for the dielectric constant is shown in Table 4.2.

Table 4.2: Validation of dielectric constant (ε_r') reconstruction using simulated data

Actual value of ε_r'	Material with ε_r''=0.01		Material with ε_r''=1.2	
	Proposed method	Analytical method	Proposed method	Analytical method
1.00	0.88	1.08	0.96	0.88
2.00	1.97	2.15	2.00	2.06
3.00	2.95	3.25	2.99	3.18
4.00	4.04	4.40	3.99	4.32
5.00	4.98	5.63	4.98	5.52
6.00	5.96	6.61	5.99	6.73
7.00	6.99	7.62	7.00	7.87
8.00	8.10	9.04	8.01	9.13
9.00	9.02	10.52	9.02	10.52
10.00	10.18	11.96	10.02	11.94

The result shows that the proposed method has an accuracy of dielectric constant reconstruction of approximately 98.0% and 99.4% for low loss (ε_r'' = 0.01) and moderate loss material (ε_r'' = 1.2), respectively. The corresponding accuracy for the analytical method is 87% for both values of the loss factor. The validation of loss factor reconstruction is shown in Table 4.3.

Table 4.3: Validation of loss factor reconstruction using simulated data

Actual value of ε_r''	Material with ε_r'=3.8		Material with ε_r'=10	
	Proposed method	Analytical method	Proposed method	Analytical method
0.10	0.12	0.11	0.13	0.12
1.00	1.00	1.05	0.97	1.28
2.00	2.00	2.08	1.98	2.56
3.00	3.00	3.08	2.98	3.84
4.00	4.01	4.04	3.97	5.15
5.00	5.03	4.97	4.97	6.49
6.00	6.05	5.84	5.98	7.89
7.00	7.06	6.64	6.98	9.37
8.00	8.09	7.33	7.99	10.98
9.00	9.11	7.84	9.01	12.73
10.00	10.13	8.05	10.02	14.61

The accuracy of loss factor reconstruction using the proposed method is approximately 97.5% and 97.0% for materials with $\varepsilon_r' = 3.8$ and $\varepsilon_r' = 10$, respectively. The corresponding accuracy using the analytical method is 93.4% and 68.8%. The measurement of loss factor $\varepsilon_r'' = 0.1$ shows large percentage of error. It shows the lowest limit of loss factor which can be measured using proposed method. It also confirms that the non-resonant method is suitable for moderate to high loss material (CON+04).

Based on validation using measured and simulated data, the proposed method has higher accuracy compared to the analytical method. Although the size of sample is already selected as suggested in [CCFD03, AFT07 and Has09], the effect of higher order modes still exists in the analytical method as can be seen in the dielectric reconstruction result of lossy material. The error becomes larger for higher loss factor. Therefore in Chapter 5, the temperature and time dependent dielectric measurement have been done with the neural network

In order to define the accuracy in the measurement, the source of error in S-parameters has been investigated. The errors in the measurement of S-parameters will propagate to errors in the measurement of dielectric properties. In order to minimize the systematical errors in dielectric properties, the source of errors in S-parameters should be investigated. The first source of error comes from the noise floor during S-parameter measurements with the vector network analyzer. The noise floor can be reduced by using intermediate frequency (IF) filtering. The PNA 5224 provides the selection of the IF bandwidth used in the measurement. However, smaller IF bandwidth will cause longer measurement time needed for sweeping over the frequency band. Between accuracy and measurement time there must be a compromise because the curing process is time-dependent. An IF bandwidth of 1 kHz was selected for monitoring the curing process. With 1 kHz IF bandwidth, the minimum time sweep for the selected frequency band is approximately 368 ms. The dynamic range for the measurement sytem is 80dB.

The second source of errors might come from the pneumatic controlled opening-closure system. However, it has been proven experimentally that the pressure closure system has only a small influence on the errors. The third source of errors comes from the sample position and orientation. The fourth error source is based on air bubbles in the resin developing during the curing process that deteriorates the sample geometrie. Therefore gas removal is applied by putting the resin in vacuum for several hours before mixing. In the next subchapter the influences of the pressure closure system, air bubbles, as well as sample orientation and position are investigated experimentally, except the influence of air bubbles since it is not possible to make air bubbles by design in the real measurement of resins. The investigation of air bubbles is performed by simulation calculations.

4.2.1 Influence of the pneumatic closure system

In waveguide-based measurements, waveguides are usually screwed to each other to ensure perfect electrical contact and to eliminate any air gaps between the two waveguide flanges. In the present setup, compressed air is selected in combination with pneumatic cylinders in order to open and close the waveguide system. As described before this allows reducing the heat transfer from the hot waveguide sample holder to the measurement set-up that may influence the system calibration.

Investigations on the influence of the pneumatic pressure to the closure system were performed as presented in Fig. 4.12 and 4.13. It can be seen that the higher the air pressure, the closer the magnitude of S_{11} to the reference value obtained when waveguides were screwed to the waveguide sample holder. The difference in phase is relatively small at various pneumatic pressures compared to the result with screwing. These facts are also found in the measurement of S_{21} (Fig. 4.13). The results show that the higher the air pressure, the closer the magnitudes of S_{21}. The difference of the S_{21} magnitude is smaller than that of S_{11}. The phase of S_{21} is almost identical for various air pressures and for the screwing system.

Reproducibility of the opening-closure system at 7 bar pressure is investigated and presented in Fig. 4.14 and 4.15.

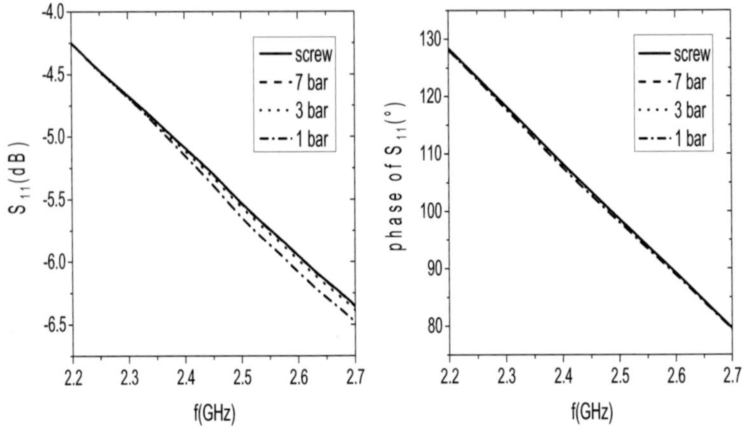

Figure 4.12: Measured S_{11} of the resin DGEBA in the crucible at different pneumatic pressure and screwing

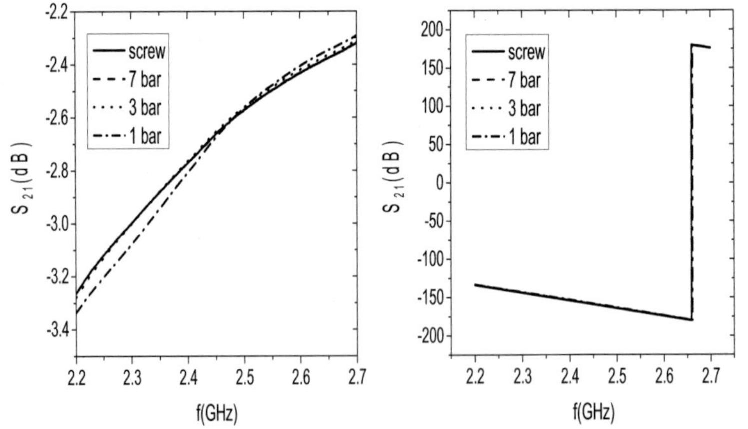

Figure 4.13: Measured S_{21} of the resin DGEBA in the crucible at different pneumatic pressure and screwing

At maximum pressure (7 bar), the reproducibility of the measured S_{11} from three measurements is very good as shown in Fig. 4.14.

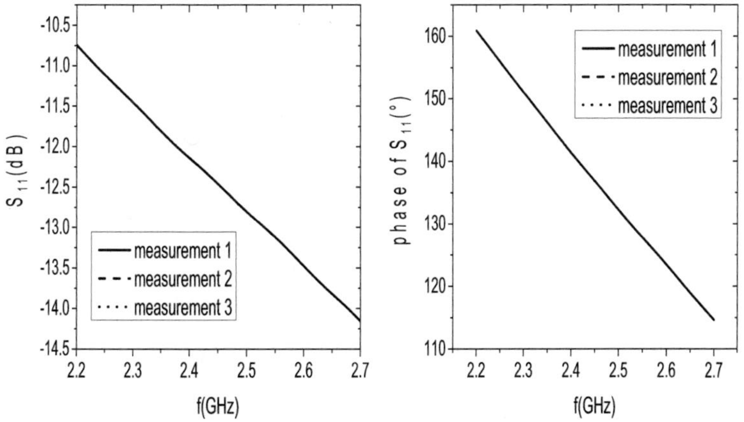

Figure 4.14: Measured magnitude and phase of S_{11} of the empty Teflon crucible at 7 bar pneumatic pressure

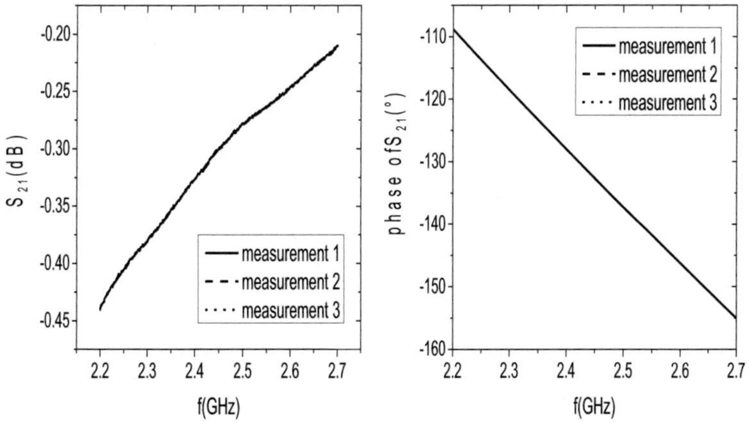

Figure 4.15: Measured magnitude and phase of S_{21} of the empty Teflon crucible at 7 bar pressure

The measured S_{21} also shows good reproducibility using the opening-closure system as can be seen in Fig. 4.15.

4.2.2 Influence of sample position

The influence of the sample position on measurement accuracy is discussed now. Experimental investigations were performed for 5r4DGEBA in a Teflon crucible. Shifting the sample from the reference position gives an additional measurement error. The error of position shifting is investigated to calculate the total errors in the measurement.

Shifting of the sample position is investigated in the experiment, as shown in the following Fig. 4.16. The shifting can be performed in *x* and *z* direction.

Figure 4.16: Teflon crucible in waveguide

The variation in magnitude and phase of S_{11} for shifting along the x direction from the center position is shown in Fig. 4.17. The results show that shifting of the sample in x direction influences both magnitude and phase of S_{11}. The variation in magnitude and phase of S_{21} for shifting along the x direction from the center position is plotted in Fig. 4.18.

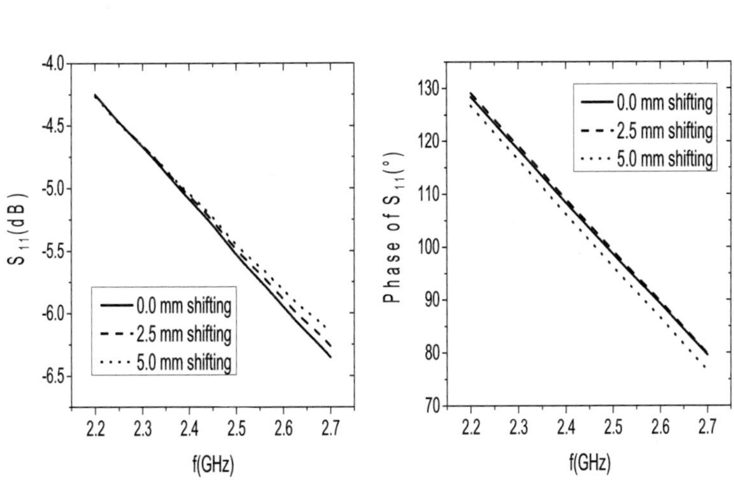

Figure 4.17: Measured magnitude and phase of S_{11} for the resin DGEBA in a Teflon crucible with different x-position

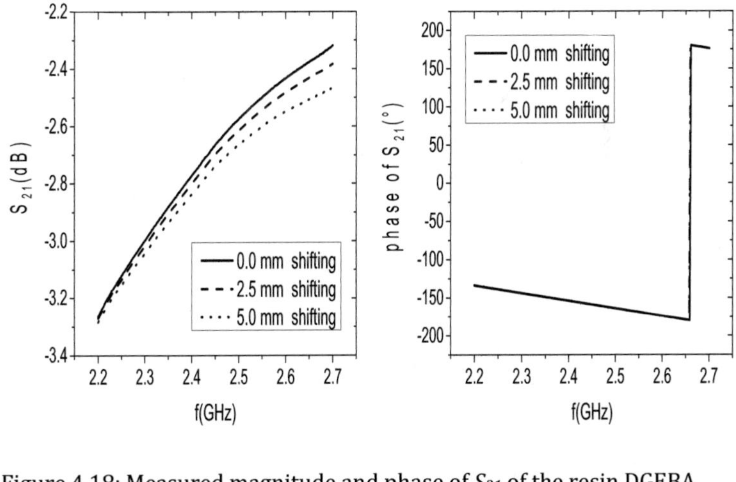

Figure 4.18: Measured magnitude and phase of S_{21} of the resin DGEBA in a Teflon crucible with different x-position

The shifting of the sample in x direction only changes the magnitude of S_{21}. The effect of shifting is larger at higher frequency both in magnitude of S_{11} and S_{21}. It can be concluded that shifting of the sample in x direction influences the measured *S*-parameters. These changes can lead to an error in the calculation of the dielectric properties.

In order to obtain detailed information about the effect of shifting, simulations in the model with different values of the dielectric properties have been performed. The dielectric properties for simulated shifting in x direction of the model are shown in Table 4.4. Both dielectric constant and loss factor are changing with the shifting. The calculation is based on the original position so that the error can be deduced. The error of the dielectric constant is around 5% and 10% for 1 mm and 2 mm shifting, respectively, whereas the error in loss factor is smaller with approximately 1% and 5 %, respectively.

Table 4.4: Simulated dielectric properties for shifting the sample in the x direction

Without shifting		1 mm shifting		2 mm shifting	
ε_r'	ε_r''	ε_r'	ε_r''	ε_r'	ε_r''
2.02	0.01	2.16	0.01	2.29	0.02
3.79	0.70	3.97	0.71	4.12	0.73
7.8	1.2	8.22	1.18	8.59	1.11

The influence of shifting in z direction is presented in Table 4.5. The error of the dielectric constant is approximately 1.5% and 3% for 1 and 2 mm shifting, respectively, whereas the error in loss factor is larger with around 5% and 10%, respectively.

Table 4.5: Simulated dielectric properties for shifting the sample in the z direction

Without shifting		1 mm shifting		2 mm shifting	
ε_r'	ε_r''	ε_r'	ε_r''	ε_r'	ε_r''
2.02	0.01	1.96	0.09	1.93	0.19
3.79	0.70	3.86	0.74	3.91	0.79
7.8	1.2	7.84	1.05	7.9	0.98

The shifting of sample positions both in x and z direction influences the accuracy of the measurement of the dielectric properties as can be seen from Table 4.4 and 4.5. The error comes from the changes of magnitude and phase of *S*-parameters that are different with the corresponding *S*-parameters at reference position. In order to reduce the possibility of this error, a special tool for placing the crucible has been developed and manufactured. The tool is used to place the crucible in the reference position so that the difference between the experiments and the model can be minimized.

In addition, the influence of sample orientation was investigated and the results are presented in Fig. 4.19 and 4.20. The orientation of the sample only results in small changes of both magnitude and phase of S_{21} and S_{11} for an orientation difference of 22.5° and 45°.

Since in the measurements such large variation of the orientation is not likely, the influence of orientation uncertainty can be neglected.

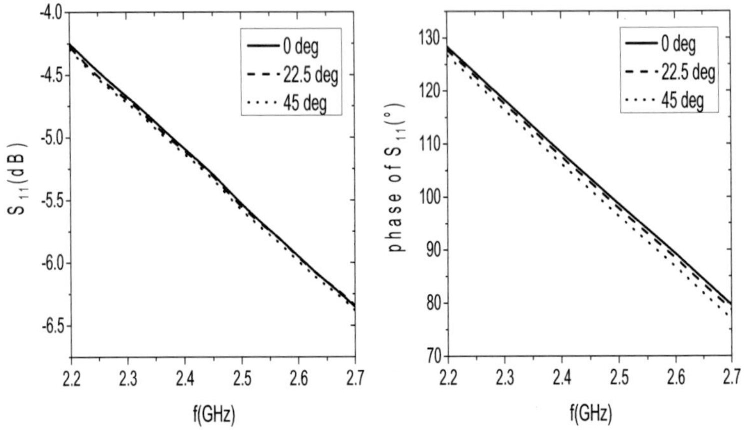

Figure 4.19: Measured S_{11} of the resin DGEBA in a Teflon crucible with different orientation (0, 22.5, and 45 deg)

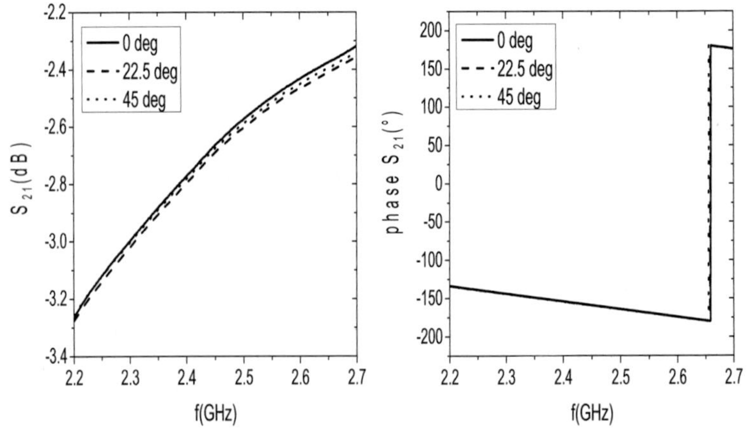

Figure 4.20: Measured S_{21} of the resin DGEBA in a Teflon crucible with different orientation (0, 22.5, and 45 deg)

4.2.3 Influence of air bubbles

Air bubbles which develop during heating and curing in the viscous resin might influence the accuracy of the measurement. The development of air bubbles can be reduced by previous gas removal using vacuum equipment.

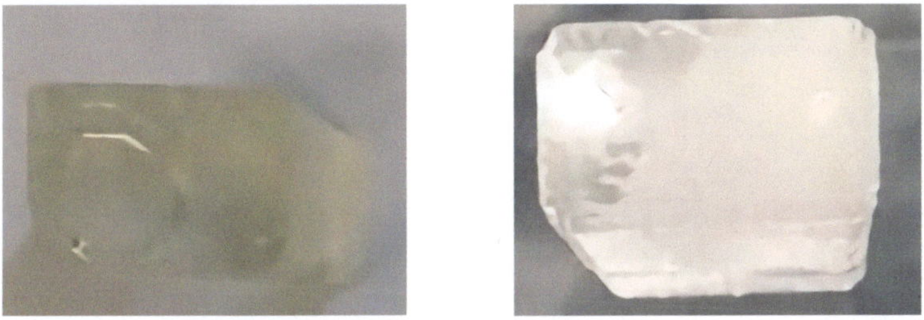

Figure 4.21: Final product of curing without (left) and with (right) gas removal

Fig. 4.21 shows the final product of curing without and with prior gas removal process. The curing without gas removal process produced a quite large air bubble in the final product which has a radius of around 6 mm. The dielectric properties slightly change due to this air bubble effect.

The air bubble only produces a small error as shown in Table 4.6. The dielectric constant and loss factor only change 0.1% and 0.2%, respectively with a 2.5 mm radius air bubble. Due to this insignificant error contribution, the effect of smaller air bubbles can be ignored. However, in the present series of experiments the gas removal process was always employed before the measurements on curing in order to minimize the air bubbles effect.

Table 4.6: Dielectric properties of sample without and with simulated air bubble

Without air buble		Air bubble with 0.5 mm radius		Air bubble with 2.5 mm radius	
ε_r'	ε_r''	ε_r'	ε_r''	ε_r'	ε_r''
2.02	0.01	2.02	0.01	2.01	0.01
3.79	0.70	3.79	0.70	3.78	0.68
7.8	1.2	7.80	1.19	7.81	1.18

4.2.4 Influence of surface level

The volume of the resin mixture in the measurements might be slightly different compared to the assumptions in the model. This condition also contributes to the error of the measurement shown in Table 4.7.

Table 4.7: Dielectric properties for different surface level of resin mixture

Actual level		-0.5 mm from the surface		-1 mm from the surface	
ε_r'	ε_r''	ε_r'	ε_r''	ε_r'	ε_r''
2.02	0.01	2.01	0.01	1.97	0.01
3.79	0.70	3.79	0.70	3.63	0.64
7.80	1.20	7.81	1.19	7.20	1.20

The surface level differences are the differences between the actual level and the reference level. Shrinking of the volume during the curing process also contributes to the error of the measurement. It can be seen in Table 4.7 that the difference in surface level of 0.5mm produces insignificant errors both in dielectric constant and loss factor which is less than 0.4% and 0.8%, respectively. However a difference of 1 mm of the surface level contributes to the error of the dielectric constant by 2%, 4% and 7% for a dielectric constant ε_r' of 2.02, 3.79 and 7.8, respectively.

4.2.5 Estimation of total error

Every source of error contributes to the total error of the measurement. In the calculation of dielectric properties where an analytical formula exists, the total error can be calculated from the derivative error formula. Since here the analytical formula for the system is very difficult to be derived, the total error is estimated based on each source of error. At least, the approximation of the worst case can be made by the following formula [Bar78]:

$$\Delta\epsilon_{tot} = \sqrt{(\Delta\epsilon_1)^2 + (\Delta\epsilon_2)^2 .. + (\Delta\epsilon_n)^2}$$

with:

$\Delta\epsilon_{tot}$ is the total error
$\Delta\epsilon_1$, $\Delta\epsilon_2$,... $\Delta\epsilon_n$ are the errors of the error sources 1, 2...up to n.

In the previous subchapter it is shown that the errors of the measurements are mainly from uncertainty of the sample position (in x and z direction) and the surface level of the resin mixture. Assuming that the maximum shift in x and z direction is 1 mm, and the uncertainty of the surface level is 0.5 mm, the total error of the dielectric constant can be calculated as:

$$\Delta\varepsilon_r' = \sqrt{(5\%)^2 + (1.5\%)^2 + (0.4\%)^2}$$

This results in a maximum error of 5.2% for the dielectric constant. In the same way the total of error of the loss factor can be calculated:

$$\Delta\varepsilon_r'' = \sqrt{(1\%)^2 + (5\%)^2 + (0.8\%)^2}$$

This results also for the loss factor in a maximum error of around 5.2%. Both values are worst case approximations so that the error most probably is lower than these values. Another way to estimate the error is by statistical approximation. The total error from all error sources can be statistically estimated by repeating the measurements on the same materials.

4.3 Materials

Polymer composites are selected as materials under test (MUT) in this study. The composites are obtained from the mixture of the resin (matrix), filler (re-enforcement) and curing agents (hardeners). The measurements are performed for the individual constituent materials and for various mixtures. The measurements on mixtures are performed during curing.

4.3.1 Epoxy resin

Epoxy resins are commonly applied in adhesives, laminates, coatings and moldings. Epoxy resins are usually obtained from chemical reaction of epichlorohydrin and a polyhydroxyl such as bisphenol A. The epoxy resin which is used in the present experiments is diglycidyl ether of bisphenol A (DGEBA), which has the molecular formula $C_{21}H_{24}O_4$ [Che]. Wide application of DGEBA is the reason why this resin is selected in this study. The chemical structure of DGEBA is shown in Fig. 4.22.

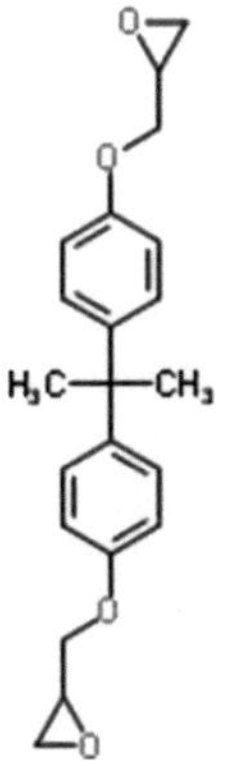

Figure 4.22: Chemical structure of DGEBA

DGEBA has different grades of viscosity, which are dependent on the molecular weight distribution. Epilox A 18-00 from Leuna GmbH is selected as the epoxy resin in this study. The characteristics of Epilox A 18-00 [Leu13] are shown in the following Table 4.8.

Table 4.8: Physical and chemical properties of Epilox A 18-00

Physical and chemical properties	Unit
Molecule mass	340.41 g/mol
Density	1.17 g/cm^3
Equivalent epoxy weight (EEW)	175-185 g/eq

The equivalent epoxy weight can be defined as the weight of the resin in gram which contains one gram equivalent of epoxy [DDRG77].

4.3.2 Filler and curing agents

Fillers are interesting parts of composites which can be used to obtain expected characteristics of the final products. Fillers can be made of biological resources, which are a part of natural composites. Natural composites are in the focus of current research because their advantage is that they can be naturally decomposed at low cost. Natural fibers are commonly extracted from plants or vegetables. Therefore, cellulose is used as filler in this study. The cellulose is also cheaper as compared to anorganic fillers. Therefore the use of cellulose will reduce the cost of producing the polymer composites. Different percentages of cellulose are investigated to understand the curing process with this filler.

Curing agents (hardeners) are also important parts of mixtures which influence the curing behavior. They also influence the final properties of mixtures. Different curing agents are investigated by starting from the slowest to the fastest one. Three kinds of curing agents are investigated: polyetheramine D230 (PEA), isophorone diamine (IPDA) and laromin (Mixture of PEA and IPDA).

The investigated curing agents are Baxxodur EC 301, Baxxodur EC 201 and Baxxodur EC 2175 produced by Baden Anline and Soda Factory (BASF). Baxxodur EC 301 is a curing agent based on PEA. The Baxxodur EC 301 a chemical structure as shown in Fig 4.23.

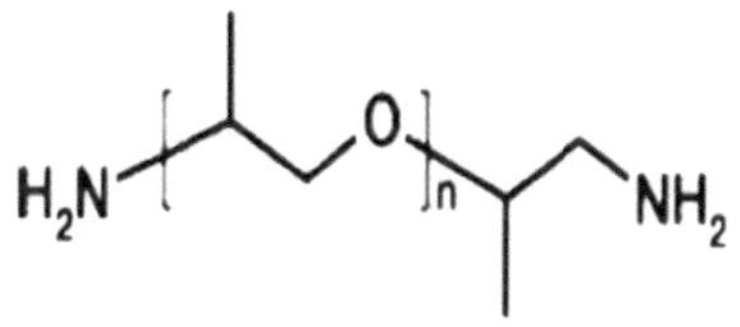

Figure 4.23: Chemical structure of Baxxodur EC301

The viscosity, density, amine hydrogen weight equivalent (AHEW), gel time, and glas transition temperature of Baxxodur EC 301 are summarized in Table 4.9 [Bas13a]:

Table 4.9: Physical and chemical properties of Baxxodur EC 301

Property	Unit	Test method
Viscosity at 23°C	9-10 mPas	DIN53015
Density	0.948 g/cm^3	ASTM D /792
AHEW	~ 61 g/eq	
Gel time (100 g at 23°C)	>550 min	ASTM D 2471
T_g	67-78°C	DSC, 10°C/min

The AHEW is the weight of the amine molecules over the number of hydrogen atoms that are attached to the nitrogen atoms. The AHEW value defines how reactive the hardener is for a certain mass of epoxy. The glass transition temperature (T_g) is the temperature of reversible transition from the solid to a rubber like phase. The glass transition temperature shows at which temperature the final polymer composite product can be applied safely.

Baxxodur EC 201 is a curing agent based on IPDA that has the chemical structure as shown in Fig. 4.24.

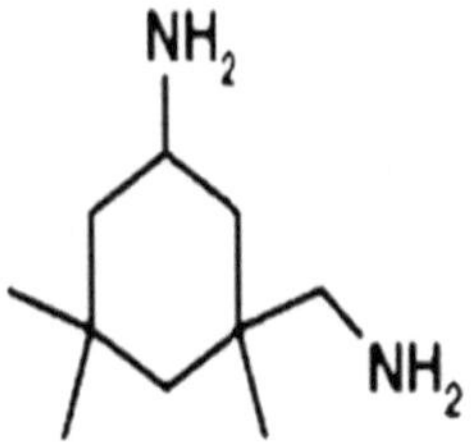

Figure 4.24: Chemical structure of Baxxodur EC 201

Table 4.10: Physical and chemical properties of Baxxodur EC 201

Property	Unit	Test method
Viscosity at 23°C	19-20 mPas	DIN53015
Density	0.924 cm³	ASTM D /792
AHEW	~ 43 g/eq	
Gel time (100 g at 23°C)	130-190 min	ASTM D 2471
T_g	128-132°C	DSC, 10°C/min

Baxxodur EC201 has high temperature resistance, high mechanical strength, excellent chemical and moisture resistance, high transparency and gives color stability to the end product. This curing agent is widely applied in structural adhesives, composite laminating, casting and encapsulation and heavy coating protection [Bas13b].

AHEW of IPDA is lower compared to PEA, showing that IPDA will react faster than PEA in the curing of epoxy (see Table 4.9 and 4.10). The glass transition of IPDA is higher than that of PEA so

4.10). The glass transition of IPDA is higher than that of PEA so that the final product of curing with IPDA will withstand better to high temperature.

Sigmacell cellulose from Sigma Aldrich has been used for the filler. The chemical structure of Sigma S3504 is shown in Fig. 4.25 and its physical and chemical properties are summarized in Table 4.11.

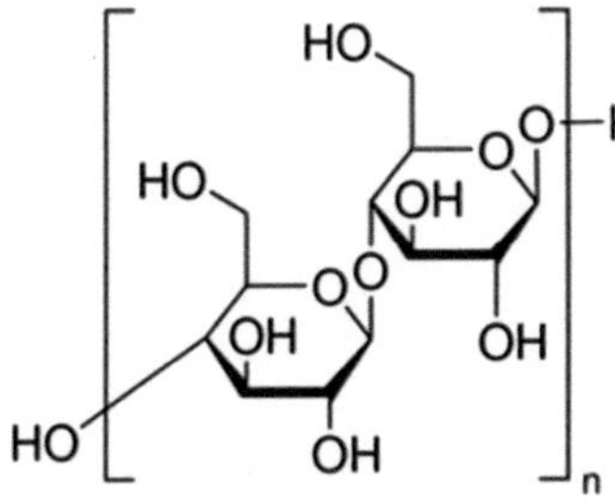

Figure 4.25: Chemical structure of Sigmacell cellulose filler

Natural fillers are commonly suitable to reinforce polymer composites due to their high strength and stiffness [BG99]. Cellulose is an essential component of all plant fibres. The S3504 consist of cellulose powder or cotton linters [Sig14].

Table 4.11: Physical and chemical properties of Sigma S3504

Property	Unit
Form	powder
Bulk density	0.6 g/cm^3
Loss on drying	1-5%
Particle size	20 μm

The complete list of materials and formulations investigated in the present study is given in Table 4.12.

Table 4.12: Formulation of the mixtures investigated in this work (phr = parts per hundred parts of resin)

Name	Formulation		
	Resin (100 phr)	Hardener (phr)	Filler Sigmacell (phr)
D-PEA	DGEBA	PEA D230 (31.6)	-
D-LAR	DGEBA	Laromin (30)	-
D-IPD	DGEBA	IPDA (23)	-
D-PEA-S30	DGEBA	PEA D230 (31.6)	30
D-PEA-S50	DGEBA	PEA D230 (31.6)	50
D-PEA-S70	DGEBA	PEA D230 (31.6)	70
D-IPD-S30	DGEBA	IPDA (23)	30
D-IPD-S50	DGEBA	IPDA (23)	50
D-IPD-S70	DGEBA	IPDA (23)	70
D-LAR-S30	DGEBA	Laromin (30)	30
D-LAR-S50	DGEBA	Laromin (30)	50
D-LAR-S70	DGEBA	Laromin (30)	70

Due to the different densities of resin and hardener, the phr (part by weight per 100 part of epoxy resin) is commonly used for mixing ratios of the materials.

The phr can be calculated with the following formula:

$$phr\ of\ curing\ agent = \frac{AHEW\ .\ 100}{EEW} \tag{4.2}$$

The formulation of the mixtures in Table 4.10 is calculated with this formula.

5 Experimental results on dielectric measurements

Results of dielectric measurements are presented in this Chapter. First, measurements on each component of the mixtures are reported. Resin, filler and curing agents are measured from room temperature up to a specific temperature. Second, the measurement results on curing are presented. Finally, the measurements of curing with various hardeners, temperatures and percentages of filler are reported.

5.1 Measurements on resin, filler and curing agents

In order to achieve a full description of the curing process with respect to the dielectric properties, besides developing a model for the reaction kinetics, the dielectric properties of each constituent need to be characterized. Therefore temperature dependent dielectric measurements of pure materials were performed.

5.1.1 Measurements on resin

Results on measurements on the dielectric constant ε_r' of DGEBA in the temperature range 25°C from to 120°C are presented in Fig. 5.1. The temperature is selected higher than 80°C in order to show the decreasing of the loss factor as effect of polarization that is explained further in Chapter 6, using the Debye model. The curve shows that the dielectric constant increases monotonically from 3.7 to 5.8 with increasing temperature.

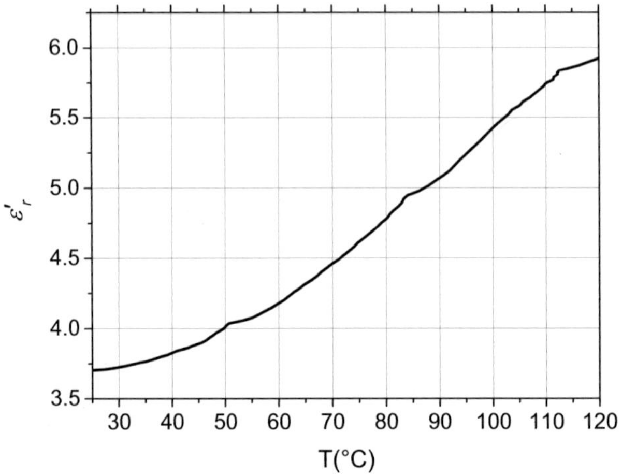

Figure 5.1: Temperature dependence of dielectric constant ε_r' of DGEBA measured at 2.45 GHz

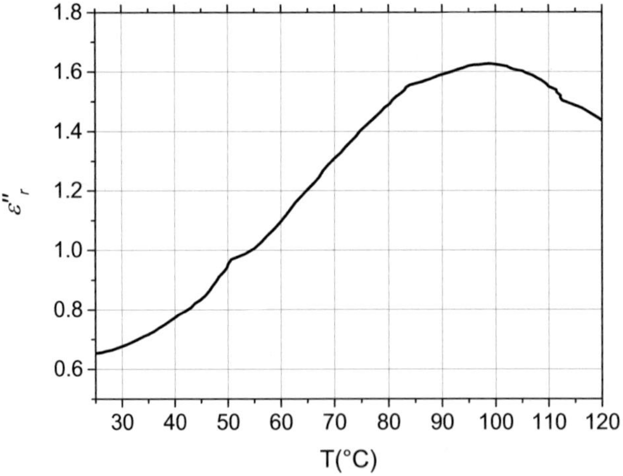

Figure 5.2: Temperature dependence of loss factor ε_r'' of DGEBA measured at 2.45 GHz

At 110°C the gradient of increasing dielectric constant becomes smaller which is an expected effect of dipole polarization as previously illustrated in the Debye model shown in Fig. 2.1.

Temperature dependent dielectric properties of DGEBA can also be found in [ZCHK04]. Their measurements were done by two methods which were using a manual or a switching system. The temperature sampling interval was between 5 - 15°C and 5 - 20°C for the manual and switching system, respectively. In comparison to this, in the measurements of the present work, temperature dependent data of DGEBA (Epilox A 18-00 with EEW 175 - 185 g/eq) is measured with a very fine, constant 1°C sampling. The absolute values of the measured dielectric constant cannot be compared directly because there is no information in [ZCKH04] on the equivalent epoxy weight of the DGEBA and the accuracy of their measurement system. The equivalent epoxy weight of DGE-BA depends on the supplier, as shown in [NBFN03, KC99, Par09]. Nevertheless the temperature dependent dielectric properties can be generally compared, because the different type of DGEBA (different EEW) does not change the temperature dependent dielectric behavior as described in [RSWZ07]. The different EEW only changes the absolute value of dielectric properties, while the common pattern of temperature dependency stays constant. The dielectric constant increased from 4.4 to 6.0 monolithically by increasing the temperature up to 115°C for the manual measurements. The result for the automatic switching system showed that the dielectric constant increased with increasing temperature but there was some deviation in some temperature points.

In order to compare the dielectric properties of DGEBA with the same EEW (175 - 185 g/eq), measurements with a coaxial probe were performed in this study at room temperature and 2.45 GHz. The resulting dielectric constant of 3.9 ± 0.3 is in good agreement with the measurement shown in Fig. 5.1. The accuracy of the coaxial probe is based on [HP97].

The loss factor ε_r" (see Fig. 5.2) reveals an increase up to 97°C, going through a maximum of about 1.6 and then starts to decrease again. In the manual measurement of [ZCKH04] the loss factor increased up to 65°C going to the maximal value of 0.8 and started to decrease again. The automatic switching system showed an increasing loss factor with oscillations up to 90°C and a decrease at 105°C. There was no measurement point between 90°C and 105°C. The pattern confirms the similar behavior compared to present study which has more measurement points and a much finer sampling to describe the temperature behavior. In addition, the fully automatic measurement system and control of temperature, as well as the method to calculate the dielectric properties improve the reliability of the measured temperature dependence of the dielectric data. The loss factor of the selected DGEBA at room temperature is 0.68 ± 0.04. The corresponding measurement with the coaxial probe shows a loss factor of 0.71 ± 0.07 which is in very good agreement.

The curing behavior of any resin strongly depends on the specific chemical mixture with hardeners, fillers and potential other constituents such as fibers for reinforcement purposes. The dielectric behavior of DGEBA significantly dominates the early stage of curing since the percentage of resin is larger compared to the hardeners. The measurement of different hardeners is presented in the following sub chapter.

5.1.2 Measurement on curing agents

Measurements on three different curing agents have been conducted. The measurements were carried out from room temperature up to 80°C. Measurements at higher temperatures are not needed since curing happens below 80°C (60°C, 70°C, 80°C). The reason to keep quite low temperature for curing is to avoid excessive exothermal reaction. The effect of the temperature on the curing process can be adequately understood from the temperature differences chosen in the present experiments. The measured temperature dependences of the dielectric constants of the three different hardeners are shown Fig. 5.3.

The dielectric constant of PEA increases from 4.5 to 5.4 with increasing temperature. Meanwhile the dielectric constant of IPDA stays almost constant at 4.3 up to approximately 45°C and then starts to decrease down to 4.1 at 80°C. Laromin combines the behavior of PEA and IPDA. The dielectric constant slightly increases from 4.4 to 4.5, and then starts decreasing at 70°C. The dielectric constant of PEA has a similar temperature behavior as DGEBA whereas the dielectric constant of IPDA has an opposite behavior.

The loss factor of PEA slightly increases from 1.4 to 1.45 up to 45°C and then starts to decrease to 1.2 with increasing temperature as shown in Fig. 5.4. The loss factor of IPDA decreases from 0.85 to 0.5 with decreasing temperature. Meanwhile the dielectric loss of Laromin stays constant at around 1.16 from room temperature to 40°C and then starts to decrease down to 0.9 with increasing temperature. The loss factor also shows that Laromin has the combined behavior of PEA and IPDA.

The temperature dependent loss factor of PEA shows similar behavior as that of DGEBA, however the change of the loss factor of PEA is smaller compared to the variation of the loss factor of DGEBA.

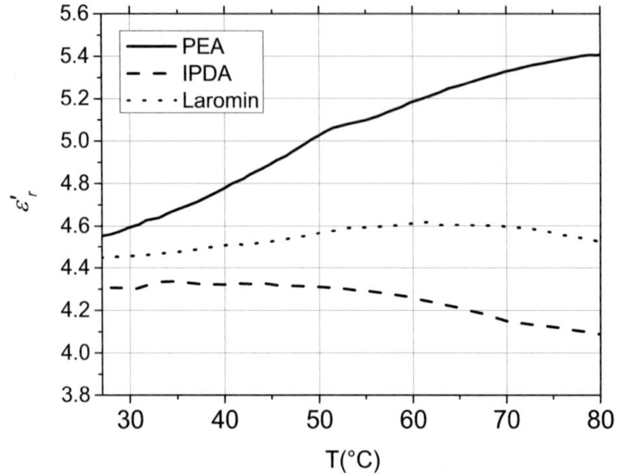

Figure 5.3: Temperature dependence of dielectric constants of three different hardeners measured at 2.45 GHz

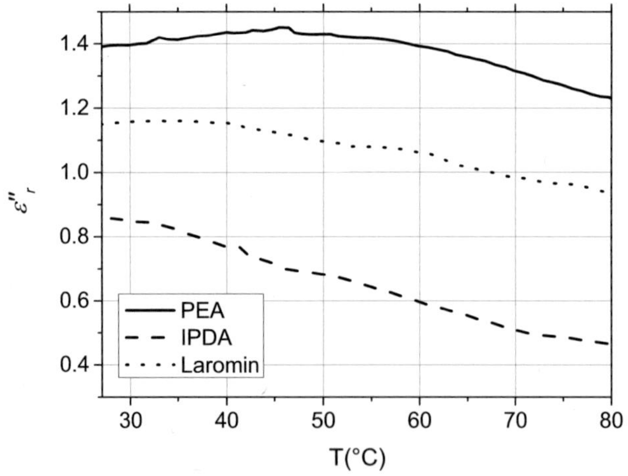

Figure 5.4: Temperature dependence of loss factor of three different hardeners measured at 2.45 GHz

The temperature dependent dielectric properties of the three hardeners could not be found in the literature. Even the dielectric measurements on these three hardeners at room temperature have not been published up to now.

5.1.3 Measurements on filler

The filler selected in this study is cellulose Sigma S3504 in powder form. Dielectric properties of powder materials are different from the intrinsic value.

Table 5.1: Dielectric properties of Sigma S3504 measured for various densities at 2.45 GHz at room temperature

Filler density (g/cm^3)	Dielectric constant	Loss factor
0.34	2.37	0.35
0.35	2.41	0.35
0.37	2.45	0.36
0.40	2.48	0.36
0.42	2.52	0.37
0.47	2.59	0.37
0.52	2.63	0.38
0.56	2.69	0.39
0.58	2.71	0.40

The intrinsic dielectric properties of Sigmacell are needed for the modelling of samples with different amount of fillers. Since only powder material is available for the measurement, the dielectric properties of Sigmacell powder have to be measured at various densities. Then the intrinsic dielectric properties can be estimated by extrapolating the dielectric property data at different densities to a sample with theoretical density (1.46 g/cm^3) [Sun05]. The results of measurement on cellulose at various densities are shown in Table 5.1. The dielectric constant and loss factor increase along with increasing density.

Measurements on different kinds of cellulose with different densities had been done in [SHA+11]. The measurement was performed by using a microwave resonator operating in the range of 700-800 MHz. The results also showed that both the dielectric constant and loss factor increased in line with increasing densities. In addition, the fitting of the measured data showed a linear increase of the dielectric properties with increasing density. Thus, the intrinsic dielectric properties of Sigmacell can be extrapolated from the measurements at various densities.

Using a fitting algorithm, the dielectric constant can be represented using a linear equation as:

$$\varepsilon_r' = 1.206d + 1.943 \tag{5.1}$$

where d is density of Sigmacell, with root mean square error (RMSE) = 0.0290 and R-square = 0.9350. The RMSE is calculated from the following formula:

$$\text{RMSE} = \sqrt{\frac{1}{N}\sum_{i=1}^{N}(y_i - \dot{y}_i)^2} \tag{5.2}$$

where i is the data number (integer from 1 to N), y_i and $\dot{y}_i$ are the measured and fitted data respectively. The coefficient of determination R^2 (R-squared) indicates how well the data fit with the model. It can be calculated from the following formula

$$R^2 = 1 - \frac{\sum_{i=1}^{N}(y_i - \dot{y}_i)^2}{\sum_{i=1}^{N}(y_i - i)^2} \tag{5.3}$$

where $\bar{y}_i$ is the mean value of the data.

The loss factor can be represented using a linear equation with RMSE = 0.0040 and R-square = 0.9480.

$$\varepsilon_r'' = 0.186\, d + 0.270 \tag{5.4}$$

Temperature dependent dielectric measurements have been performed in the present work up to the maximal density of 0.58 g/cm³. The temperature dependence of dielectric constant and loss factor at 2.45 GHz is plotted in Fig. 5.5 and 5.6, respectively. Both the dielectric constant and loss factor increase almost linearly with increasing temperature. The small fluctuations are within the predicted uncertainty of the system.

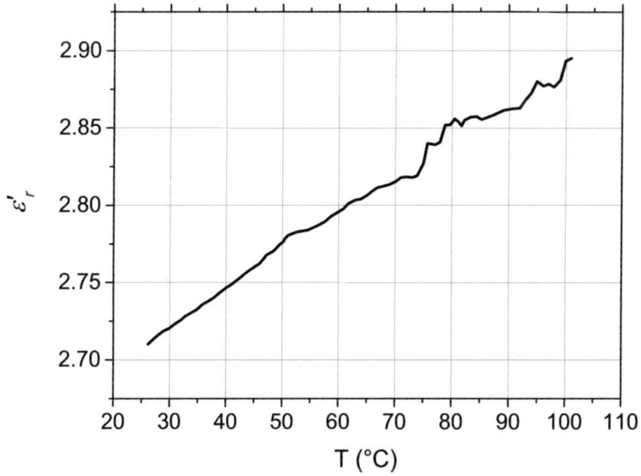

Figure 5.5: Temperature dependence of dielectric constant of Sigma S3504 measured at 2.45 GHz

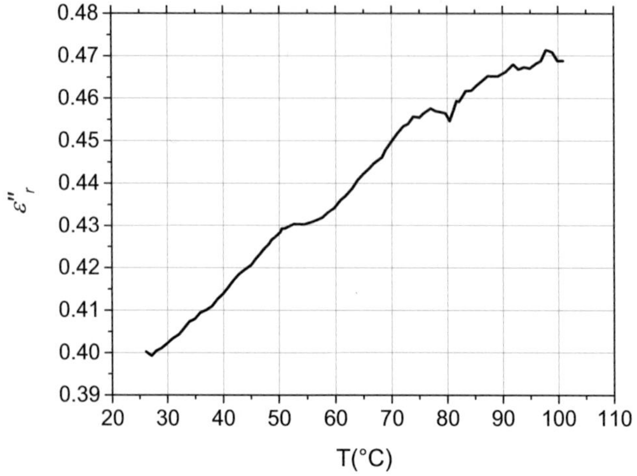

Figure 5.6: Temperature dependence of loss factor of Sigma S3504 measured at 2.45 GHz

These measurements are original and cannot be found in published literature. In [YKK09] the dielectric constant of cellulose electro-active paper had been measured at low frequencies (100 Hz – 1 MHz). There the dielectric constant also increases with increasing temperature.

5.2 Measurements on curing

Curing of a resin can be performed at room temperature using curing agents as an accelerator for initiating cross-linking reactions. However, this cold curing usually does not produce complete cross-linking. Post-curing at an elevated temperature is usually needed for better cross-linking. In practice the optimal post curing temperature lies in the range of the maximum glass transition temperature of the particular thermosetting resin. The

T_g of thermosetting resins depends on the hardener which is mixed in to the mixture. In [Ehr10] hot curing at a certain elevated temperature is recommended for obtaining highly cross-linked thermosetting resins. In this case specific curing agents that only are activated at certain elevated temperatures should be applied. This means that the storing of a mixture for limited time at room temperature is possible. Then, the preparation of the mixture for characterization of curing is possible because of the extreme low cross-linking reaction at room temperature for this limited time.

The predetermined temperature schedule of curing can be stepwise or at constant temperature. Stepwise heating is commonly used in order to get a good curing result. Stepwise curing is often carried out to control temperature dependent shrinkage and the subsequent stresses [Ehr10]. However, the main goal of the present work is to characterize the curing based on dependency on temperature, amount of filler, and kind of hardener. Stepwise heating causes difficulties in the analysis of the influence of different temperatures to the curing. Furthermore, the effect of either curing agent or filler cannot clearly be understood with the stepwise heating. So the best way to investigate the influence of different temperatures, curing agents and percentages of filler is by applying a constant curing temperature. However, as already mentioned, in real curing processes, it is impossible to obtain a desired constant temperature from the very beginning that means within zero seconds.

The preheating stage has to be considered for accurate curing monitoring. This fact leads to the scheme chosen in this work. The temperature is increased with a maximal rate of 1°C/minute from room temperature up to a specific curing temperature, and then the curing temperature is kept constant till the end of the curing process as shown in Fig. 5.7 and 5.8.

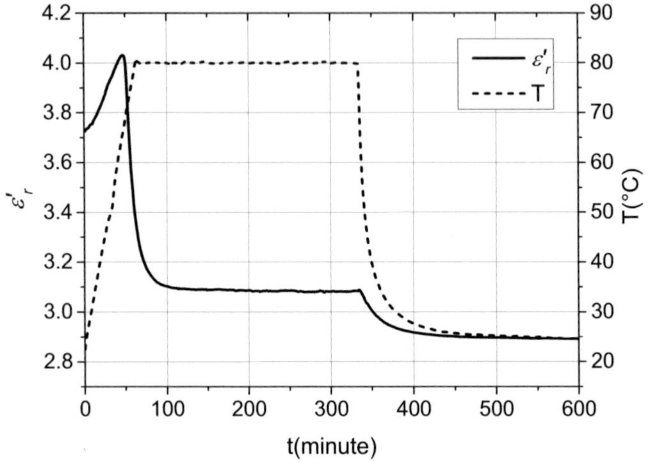

Figure 5.7: Dielectric constant of D-IPD measured at 2.45 GHz during 80°C curing and cooling

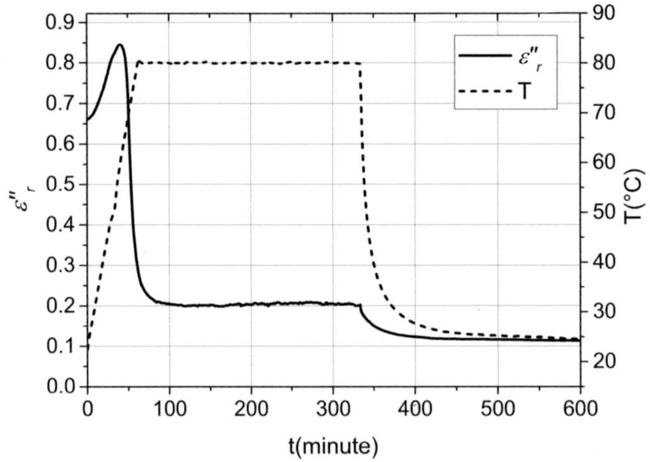

Figure 5.8: Loss factor of D-IPD measured at 2.45 GHz during 80°C curing and cooling

The cross-linking reaction depends on the reactivity of the hardener, the reaction temperature, and the ratio of epoxy equivalent mole to EEW value [DHC+99]. Therefore, in order to investigate one parameter, the other parameters have to be kept constant.

Dielectric measurement on analyzing the degree of curing had been done for low frequencies empoying the commercial instrument DEA 2970 (TA Instrument) at 0.1 Hz – 10 MHz in [KC99]. The time dependent dielectric measurement was not published; instead, the time dependent degree of curing was presented. Other measurements at the low frequency of 1 kHz can be found in [HJPK13]. The absolute values of dielectric properties were not published there; instead the normalized logarithm of dielectric properties was displayed. Measurements of dielectric properties of mixtures at 2.45 GHz during curing are conducted in the present work to inspect the behavior of curing due to different curing temperatures, filler percentages and various hardeners. In the Fig. 5.7 a measurement of the dielectric constant during the curing and cooling processes is presented. The cooling stage is measured in order to investigate the further change of the dielectric properties during cooling. Furthermore, the temperature dependent dielectric data of the final product can also be obtained from the measurements on the cooling stage.

The mixture is heated up from room temperature (24°C) up to 80°C empolying a heating rate of 1°C/minutes. The measured temperature rise is in good approximation linear during the heating process for achieving 80°C. The system is also able to keep a constant curing temperature at 80°C with relatively small fluctuations (± 0.1°C). This selected temperature scheme is much closer to the real curing temperature scheme compared to the investigations in [BSEB93, KC99 and HJPK13]. In these publications, isotherm curing was assumed starting from time zero at the beginning of the characterization. The constant temperature scheme was implemented by ignoring the heating period to the constant curing temperature. The constant scheme makes the modelling of the dielectric properties easier. The ramp-constant temperature scheme employed in the present work is close to the real curing. The curing process surely requires time to achieve a constant temperature. In this case the modelling of the dielectric, the heating period has to be included. In the literature, no model is found for this scheme.

In Fig. 5.7, the dielectric constant follows the temperature pattern slightly increasing up to about 4.0. Before reaching a constant temperature the dielectric constant goes down. This indicates that the cross-linking reaction goes faster. The increase of the dielectric constant of the mixture has the same pattern as those for each of the components. This means that the temperature dependent properties of the mixture are the combination of temperature dependent properties of each component.

The corresponding measurements on the loss factor are shown in Fig. 5.8. The loss factor also displays the same pattern behavior during curing. The loss factor increases with increasing temperature and then falls exponentially after reaching of a certain temperature. Finally the loss factor goes to a constant value when the mixture has been heated for a certain time at a constant temperature. In the cooling stage, both dielectric constant and loss factor decrease monotonically along with the decreasing temperature. The dielectric properties then converge to a constant value when the temperature reaches again room temperature. This behavior is reproducible indicating that there is no further curing reaction.

The dielectric behavior at 2.45 GHz during curing with increasing temperature from room temperature to curing temperature could not be found in the literature. Combined ideal constant heating and measurement of the dielectric properties at 2.45 GHz can be found in [MKB92 and NND04]. While the dielectric properties at low frequency can be found in [BSEB93, KC99 and HJPK13]. Only the loss factor had been used for monitoring of curing in these publications. In comparison, the novel curing monitoring techniques in the present work uses both dielectric constant and loss factor for curing characterization.

5.2.1 Measurement at different curing temperatures

Curing of all mixtures was measured at three different temperatures to study the influence of the curing temperature. The measurements have been done for three different mixtures. The preset temperature for each mixture is selected so that all the curing temperatures were reached in approximately the same time. The heating rates are around 1°C/min, 0.83°C/min and 0.67°C/min for 80°, 70° and 60°C curing temperatures, respectively.

The influence of the different curing temperatures is clearly presented in Fig. 5.9. The peak of the dielectric constant for the 80°C process is the largest one followed by the 70°C and 60°C values, respectively. The dielectric constant peak of the 80°C scheme is 4.6 and was reached after 52.2 minutes at 73.9°C. Meanwhile the 70°C scheme reaches the peak value of 4.4 after 65.9 minutes at 67.6°C. The 60°C scheme reaches the peak value of 4.2 after 62.5 minutes at 60.0°C.

This fact is expected based on the combination behavior of each constituent. As it can be previously seen in Fig. 5.2 and 5.3, the dielectric constant of DGEBA and PEA increase during increasing temperature. The mixture has the combined behavior of the constituent which has been confirmed by the results shown in Fig. 5.9.

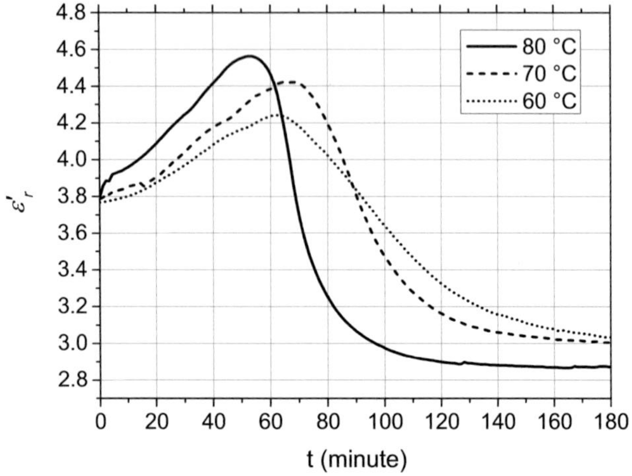

Figure 5.9: Dielectric constant of D-PEA measured at 2.45 GHz during curing with different temperatures

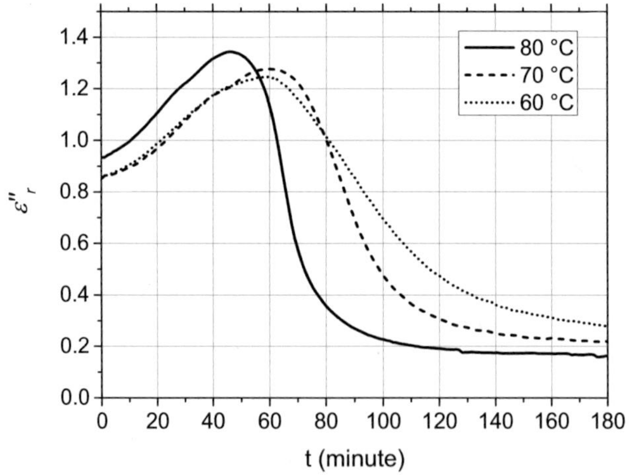

Figure 5.10: Loss factor of D-PEA measured at 2.45 GHz during curing with different temperatures

The peak of each scheme does not reach the maximum curing temperature except for the 60°C scheme because the effect of the cross-linking reaction decreases the value of the dielectric constant. The results show that the dielectric constant is increasing with rising temperature.

This indicates a slow cross-linking reaction in the early stage of curing. The dielectric constant measured at 80°C falls exponentially during the heating time from 50 to 100 minutes, and then converges to a constant value at 150 minutes. In order to make the real curing more efficient, the process can be stopped when this constant value is reached. The fact that the dielectric constant converges to a certain constant value shows that there is no more curing reaction. However, the desired value of the dielectric constant has to be defined before the real curing process. Furthermore, this information is also useful for a stepwise curing process. The procedure of stepwise heating can be arranged based on the behavior of the dielectric properties. This means that the temperature has to be increased for initiating further cross-linking reaction. The characterization of every mixture before the real curing will lead to the efficient use of energy. After reaching the peak value, the dielectric constant drops exponentially with different slopes.

The loss factor pattern shows the same behavior as the dielectric constant pattern (Fig. 5.10). Some discrepancy for the loss factor is found at the beginning of curing because of different initial room temperature. The final loss factor also clearly shows that the measurements at 70°C and 60°C do not produce 100% cured material. The steady state value for the 80°C curing process is the lowest one whereas that of the 60°C process is the highest. This fact indicates that the degree of curing of the material is different for different curing temperatures. Dielectric properties of D-PEA cannot be found in the literature.

Since the slopes in the shown permittivity data represent the reaction kinetics of the polymerisation process at the different curing temperatures, the time derivatives of the dielectric properties are presented in Fig. 5.11 and 5.12.

The sharpest downward slopes of the dielectric constant for the curing temperatures 80°C, 70°C and 60°C are around 0.084/min, 0.043/min and 0.021/min, respectively. The higher the temperature the faster the constant value is reached. The loss factor also shows that the curing at 80°C has the sharpest down slopes followed by 70°C and 60°C processes which are around 0.060/min, 0.035/min and 0.018/min, respectively.

The temperature dependent permittivity of cured materials with certain degree of curing had been published in [ZCHK04]. The results showed that for a similar degree of curing, both the dielectric constant and the loss factor increase with increasing curing temperature. Investigations on DGEBA using the hardener diaminocyclohexane (DCH) had been presented in [RBF05]. The measurement was done using the commercial DEA 2970 instrument at 0.1Hz – 100 kHz. The result showed that the loss factor decreased exponentially during isothermal 70°C curing. Characterization of the curing state of DGEBA with diaminodiphenil sulfone (DDS) had been presented in [WMS02]. The loss tangent decreases by increasing degree of curing so that it can be concluded that the loss tangent decreases with the time of curing. In comparison, in the present work, the detailed behavior of both dielectric constant and loss factor is presented.

As already mentioned earlier the goals of this study are both monitoring of curing and providing dielectric properties at 2.45 GHz of some technically important mixtures during curing. The knowledge of the dielectric property data during curing is the prerequisite for designing a suited microwave assisted system for effective curing.

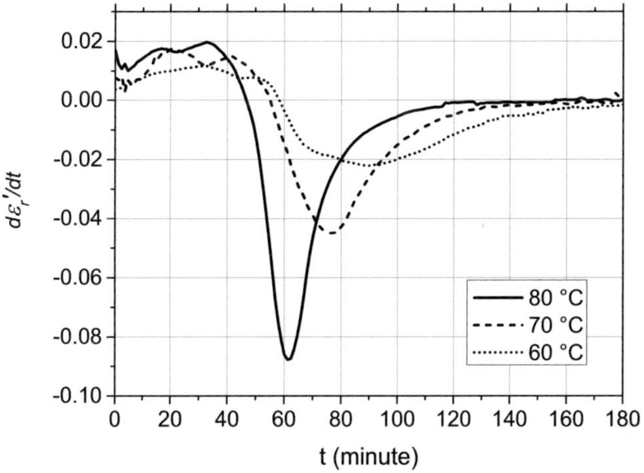

Figure 5.11: Time derivative of dielectric constant of D-PEA measured at 2.45 GHz during curing with different temperatures

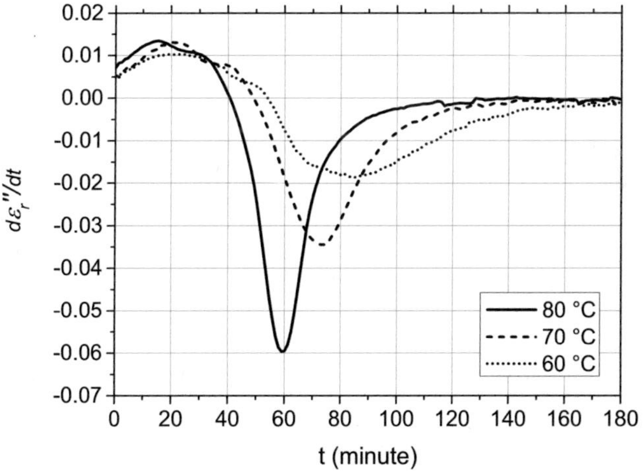

Figure 5.12: Time derivative of loss factor of D-PEA measured at 2.45 GHz during curing with different temperatures

Based on the measurement results of D-PEA at different curing temperatures, both dielectric constant and loss factor can be used for curing monitoring at 2.45 GHz. A dielectric analyzer is commercially available at low frequency below 1 MHz. The monitoring of curing is only based on the loss factor due to the significant error of the measurements of the dielectric constant at and below the megahertz frequency range [SC83]. The problem does not exist at 2.45 GHz as is proven by the following figure:

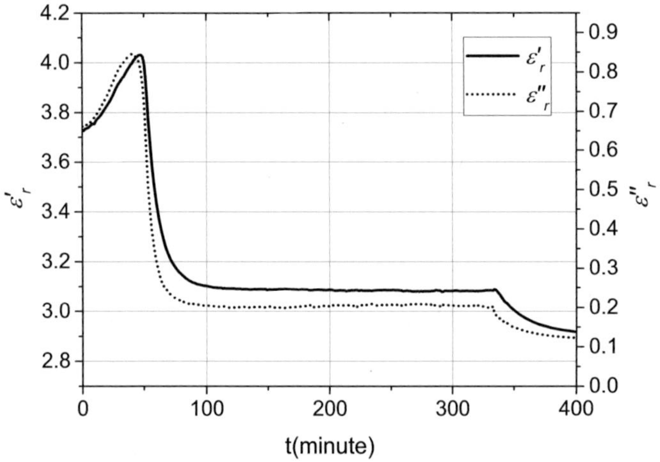

Figure 5.13: Dielectric properties of D-IPD mesuared at 2.45 GHz during 80°C curing and cooling stage

The patterns of dielectric constant and loss factor are similar during the curing and cooling stages. Both dielectric constant and loss factor are used in the modelling of the degree of curing in Chapter 6. By using both properties, the uncertainty in the degree of curing caused by measurement errors can be minimized. The common modelling of the degree of curing published in literature is only based on the loss factor [HJPK13, RBF05]

For modelling the curing process and the degree of curing based on permittivity measurements, the dielectric data and its temperature dependence should be known for the uncured as well as cured materials. Therefore, measurements of the cured material during the cooling stage have been performed in this work. The temperature dependent dielectric constant of the fully cured D-PEA is presented in Fig. 5.14. The dielectric constant increases logarithmically with increasing temperature. The temperature dependent loss factor of cured D-PEA is shown in Fig. 5.15.

The loss factor of cured D-PEA increases with increasing temperature. This behavior is similar to the loss factor during curing at low frequency in the 1-13 MHz range [ADA09]. There a mixture of DGEBA with triethylene tetramine was measured using a parallel plate capacitor.

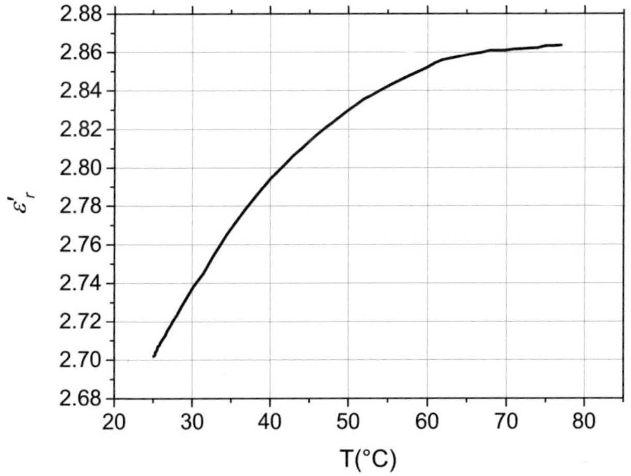

Figure 5.14: Dielectric constant of cured D-PEA measured at 2.45GHz

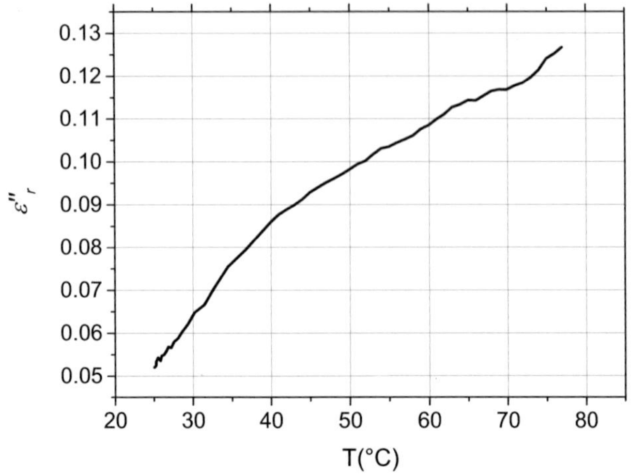

Figure 5.15: Loss factor of cured D-PEA measured at 2.45 GHz

5.2.2 Measurements with different curing agents

Influence of different hardeners is investigated in this subchapter. In order to separate the effect of hardeners from different temperature effects, the heating rate and curing temperature is kept constant at approximately the same value for all mixtures. The measurements have been done for mixtures with different hardeners at the curing temperature of 80°C with a constant heating rate of around 1°C/minute. The measured dielectric constants of mixtures with different hardener are shown in Fig. 5.16. The maximum of the dielectric constants is different for every mixture. This is just as expected because of the effect of temperature dependent behavior of the different curing agents. The dielectric constant of PEA increases from room temperature up to about 60°C, while the dielectric constant of IPDA increases up to 50°C and then decreases slightly.

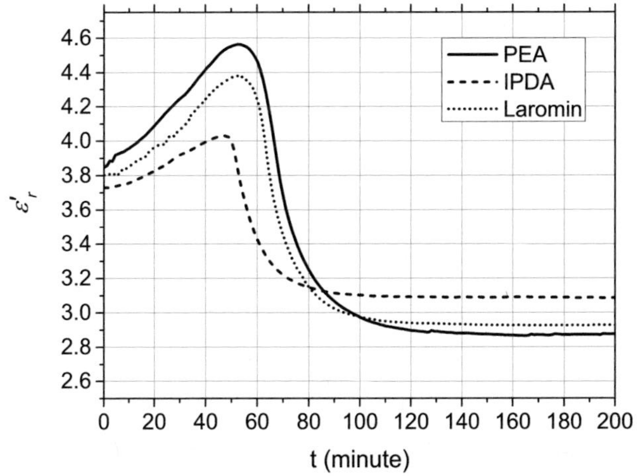

Figure 5.16: Dielectric constant of DGEBA with different hardeners measured at 2.45 GHz during 80°C curing

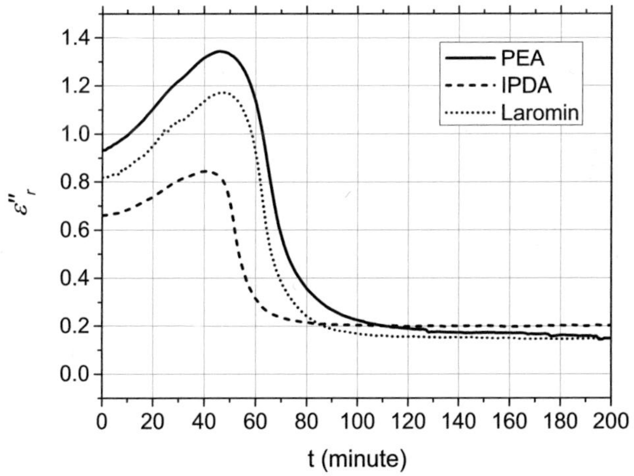

Figure 5.17: Loss factor of DGEBA with different hardeners measured at 2.45 GHz during 80°C curing

The dielectric constant of Laromin decreases from about 55°C. So the achievable dielectric constant of PEA is the highest one followed by Laromin and IPDA which are approximately 4.6, 4.4 and 4.0, respectively. Furthermore, based on the properties summarized in Tables 4.9 and 4.10, IPDA is faster than PEA, and Laromin which is the combination of the two curing agents has an intermediate characteristisc. IPDA reaches a constant value of the dielectric constant faster as compared to Laromin and PEA which is about after 35, 50 and 65 minutes from the time when the peak value was achieved, respectively.

PEA converges to the lowest value of the dielectric constant compared to Laromin and IPDA. The measurement of the mixtures at the beginning of curing (Fig. 5.16.) shows the opposite difference of the dielectric constants. According to the dielectric mixing rule, the dielectric constant of the final product of curing should have the same pattern of difference as at the beginning of curing. The reverse difference of the dielectric constants indicates that the mixture of IPDA and Laromin are not fully cured. This indication is confirmed by the DSC measurement in Table 5.2. The degrees of curing for DGEBA with PEA, Laromin and IPDA are 100%, 95.8% and 89.53%, respectively. The loss factor of DGEBA with three different hardeners is plotted in Fig. 5.17. It has the same temperature behavior as the dielectric constant. The maximum value of the loss factor of PEA is also the highest one compared to Laromin and IPDA, namely approximately 1.34, 1.17 and 0.85, respectively.

The loss factor of IPDA is almost constant after around 100 minutes, while those of Laromin and PEA reach an almost constant value at 150 and 200 minutes, respectively. This information is useful for the optimization of stepwise curing in a real curing process. Shorter curing times can be decided from this information.

The small variations of dielectric constants and loss factors indicate that again the temperature is not sufficiently high to trigger the further curing reaction. In order to obtain an efficient process in real curing, the curing temperature has to be increased when the dielectric constant and loss factor reach the constant value.

The loss factor at the final stage has different behavior compared with the dielectric constant. The loss factor of PEA is in the middle between that one of IPDA and Laromin and finally approaches to the loss factor of Laromin at the end of the curing. The reason for this is that the differences of dielectric constants at the beginning of curing are smaller compared to those of the loss factors. Based on Table 5.2, D-PEA is fully cured so that the dielectric constant and loss factor go down to small values The fully cured condition makes the dielectric constant of D-PEA smaller than that of D-Lar because of the small difference of the dielectric constants which is 0.02 (normalized value). However, this condition does not make the loss factor of D-PEA smaller compared to that of D-Lar because the normalized difference is much larger (0.18). This is also confirmed by the measurement of pure curing agents (Fig. 5.4). This fact proves that the additional measurement of the dielectric constant improves the accuracy of curing monitoring. At least the curing stage can be easier decided for the dielectric constant without further analysis.

In order to see the slopes of the dielectric properties during curing with different curing agents, time derivative dielectric properties are plotted in Fig. 5 18 and 5.19.

Fig. 5.18 shows that the maximal downslopes of dielectric constants are 0.084/min, 0.082/min and 0.06/min for PEA, Laromin and IPDA, respectively. The maximal downslopes of the loss factors are 0.060/min, 0.062/ min and 0.045/ min for PEA, Laromin and IPDA, respectively.

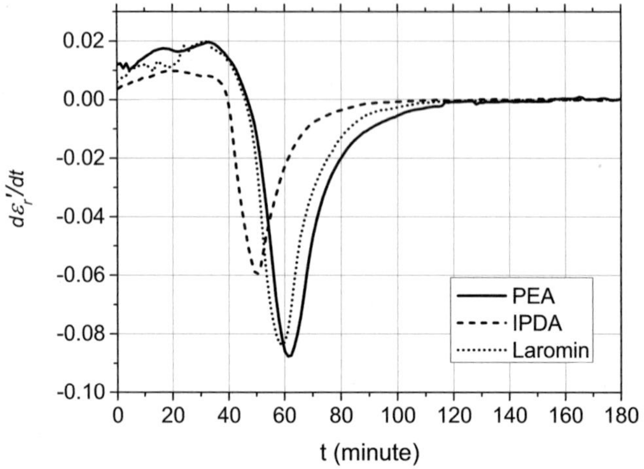

Figure 5.18: Time derivative of dielectric constant of DGEBA with different hardeners at 2.45 GHz during 80°C curing

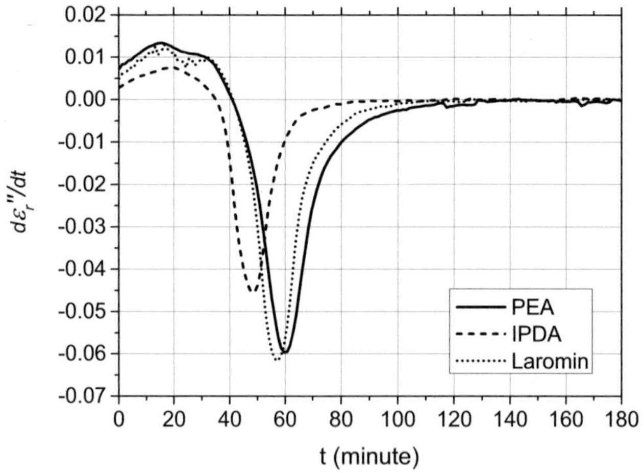

Figure 5.19: Time derivative of loss factor of DGEBA with different hardeners at 2.45 GHz during 80°C curing

The difference of the slopes between dielectric constant and loss factor is also caused by the different normalized values of dielectric properties of the curing agents as already explained in the case of the difference of the final dielectric properties.

Measurements of dielectric properties during curing with different curing agents cannot be found in the literature. Nevertheless the investigation of curing with one curing agent was published in [TOH+08]. Different mixtures were investigated during curing using an electrode arrangement. The curing behavior was represented by the amount of charge near anode and cathode. The measurement of the charge near the anode showed that the charge decreases exponentially with increasing curing time with different exponential slopes for the different mixtures. However, the validation of the degree of curing was not performed in this publication [TOH+08].

5.2.3 Measurements with different amount of filler

Measurements using three different weight percentages of filler where done for each mixture of DGEBA and hardener. The dielectric constant of three different amounts of filler and one without filler is shown in Fig. 5.20. The dielectric constants of mixtures of D-LAR with different percentages of Sigmacell filler measured during heating are relatively similar: 3.82, 3.83, 3.85 and 3.87 for 70 phr, 50 phr, 30 phr and 0 phr of filler, respectively. The filler makes the cross-linking reaction slower as indicated by the different slopes for different amount of filler.

The dielectric constant of the final product clearly shows that the lower the percentage of filler, the higher the degree of achievable curing at the same curing temperature. This conclusion is supported by the DSC measurement in Table 5.2.

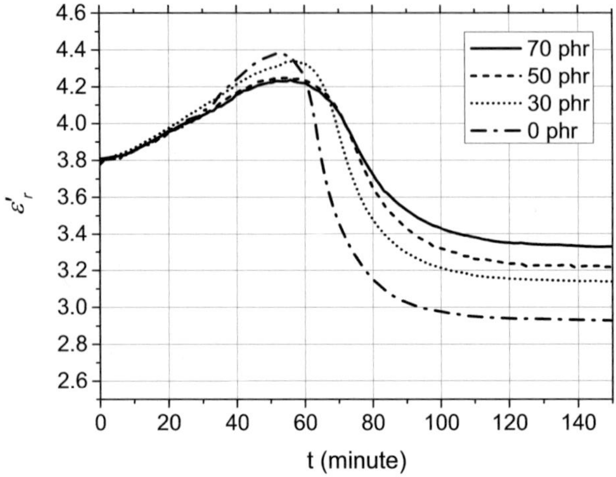

Figure 5.20: Dielectric constant of D-LAR measured at 2.45 GHz for different amounts of filler during 80°C curing

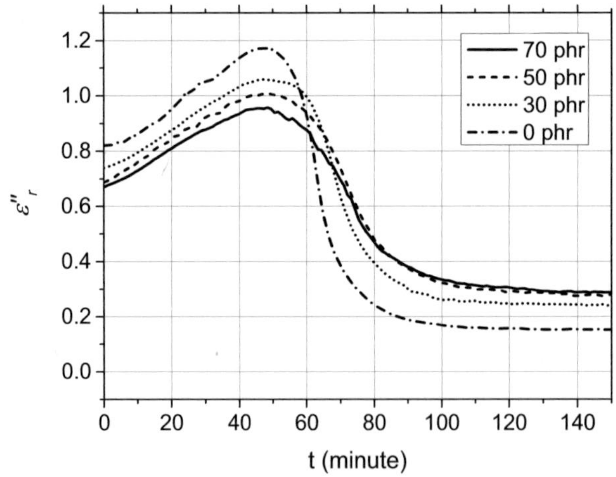

Figure 5.21: Loss factor of D-LAR measured at 2.45 GHz for different amounts of filler during 80°C curing

The degrees of curing of the mixtures with 70 phr, 50 phr, 30 phr and 0 phr of filler are 93.53%, 94.22%, 94.51% and 95.80%, respectively. This information reveals the fact that the temperature should be increased according to the percentage of filler to achieve an equivalent and sufficient degree of curing. The loss factor is slightly different at the beginning of curing (see Fig. 5.21) as expected from the mixing rule formula. After about 50 min the loss factor of the D-LAR formulation without filler is the highest one followed by the formulations blended with 30 phr, 50 phr and 70 phr of Sigmacell, respectively. The different loss factors also show that the filler decreases the speed of curing. The loss factor of the final product also shows that the degree of curing is decreasing by increasing the percentage of filler. In order to see the slopes of the dielectric properties, time derivatives of the dielectric properties of curing are presented in Fig. 5.22 and 5.23.

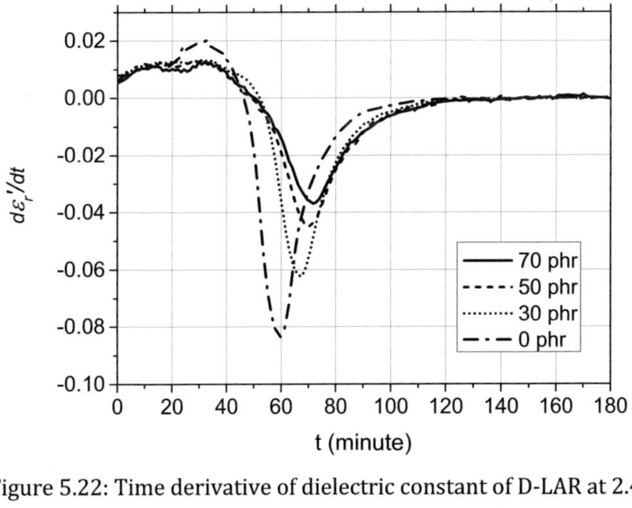

Figure 5.22: Time derivative of dielectric constant of D-LAR at 2.45 GHz for different amounts of filler during 80°C curing

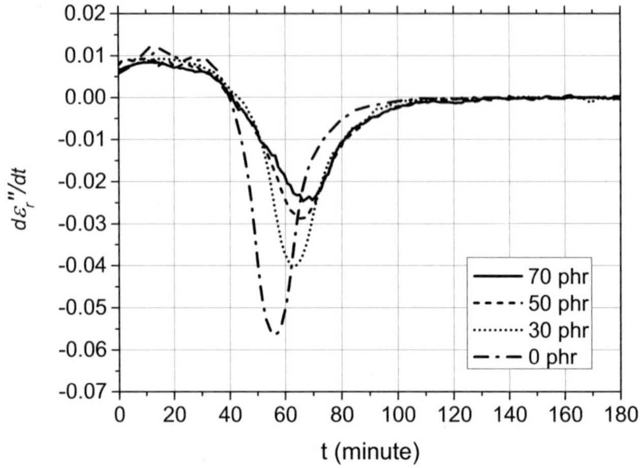

Figure 5.23: Time derivative of loss factor of D-LAR at 2.45 GHz for different amounts of filler during 80°C curing

The sharpest slopes of the dielectric constant of mixtures with 0 phr, 30 phr, 50 phr and 70 phr of filler are approximately 0.084/min, 0.062/min, 0.044/min and 0.036/min, respectively. The corresponding sharpest slopes of the loss factor are 0.066/min, 0.040/min, 0.028/min and 0.025/min, respectively. The sharpest slopes is reached earlier for the the mixture with lower amount of filler. It also gives indication that the greater the percentage of filler, the slower is the reaction rate.

5.3 Curing state of processed samples

In order to investigate the accuracy of the proposed technique for monitoring of curing, the final curing product has been characterized with Differential Scanning Calorimetry (DSC). The degree of curing derived from the measurements of dielectric properties is explained in Chapter 6.

Table 5.2: DSC measurements of cured materials

Material	Curing temperature (°C)	ΔH (J/g)	Rest energy (J/g)	Degree of curing (%)
D-PEA	60	502.52	43.34	91.75
D-IPD	70	502.52	15.02	97.01
D-LAR	80	502.52	7.67	100
D-LAR-S30	80	546.36	57.21	89.53
	80	549.29	23.06	95.80
D-LAR-S50	80	423.43	27.64	93.53
	80	427.09	24.49	94.22
D-LAR-S70	80	336.21	18.46	94.51

The given transition enthalpy ΔH of D-PEA is the average value of the three cured materials. The accuracy of the measurements is approximately ± 5% [Tho].

6 Modelling and discussion of obtained dielectric properties

This Chapter describes and discusses the novel modelling scheme for obtaining the dielectric properties of mixtures during curing. It starts with the investigation of mixing rule models for mixtures. Then, the model of temperature dependent dielectric properties of resin and curing agents is explained. Finally a novel model of curing is developed and compared to the experimental results.

6.1 Dielectric mixing rule models

The best suited dielectric mixing rule has to be choosen before modelling the dielectric data. The addition of the hardener to the mixture starts certainly cross-linking reaction after a certain time of mixing which might change the dielectric properties of the mixture. Meanwhile, there is no chemical reaction for mixing with fillers. The choice of the suited dielectric mixing rule of resin and hardeners was performed at the beginning of the curing process so that the influence of the curing reaction on the mixing rule can be neglected.

Dielectric properties of DGEBA blended with different hardeners were calculated with the LR and MG mixing formulas (2.14 and 2.15) and compared to the measured results as shown in Table 6.1. The dielectric constant of D-PEA calculated using the LR law lies in the range between 3.93 and 3.96 for ς in the range of -1 to 1. The result shows very good agreement between calculation and measurement. The dielectric constants calculated with the LR formula are in good agreement with the measurement results.

Table 6.1: Measured and theoretical dielectric properties calculated using different mixing rules for resin and hardeners

Material	Measurement		LR law with $\varsigma = -1$		MG law	
	ε_r'	ε_r''	ε_r'	ε_r''	ε_r'	ε_r''
DGEBA	3.73	0.68				
PEA	4.56	1.40				
IPDA	4.30	0.85				
Laromin	4.40	1.15	4.48	1.17	4.43	1.11
D-PEA	3.92	0.90	3.93	0.88	4.05	0.90
D-IPD	3.77	0.73	3.84	0.72	3.91	0.73
D-LAR	3.89	0.83	3.90	0.81	3.99	0.83

The differences in dielectric constant between measurements and estimated results by LR law are 0.01 for D-PEA and D-LAR, and 0.07 for D-IPD. The difference for D-IPD may result from the slight decrease of the dielectric constant by curing at already 30°C which is around 0.07% for D-PEA and D-LAR and 0.15% for D-IPD, respectively. So the curing degree of D-IPD at 30°C is twice as large as compared to the other mixtures. The difference of the loss factor calculated by the LR law is around 0.02 compared to the measurement results. The measured and calculated dielectric properties of all the mixtures are consistent with the calculations using the LR formula. D-PEA has the largest values of dielectric constant and loss factor followed by D-LAR and D-IPD. This is an expected result because pure PEA has the largest values compared to the other hardeners.

The consistency of the LR mixing rule is also confirmed by the dielectric properties of D-LAR which have intermediate values between D-PEA and D-IPD. The MG mixing law also gives consistent result as shown in Table 6.1.

The measured and estimated dielectric properties of resin mixed with Laromin hardener and different amouts of filler is shown in Table 6.2. In order to see the effect of filler to the mixing, the resin and hardener is kept constant in its weight percentage. The intrinsic dielectric properties of Sigmacell are obtained by extrapolation of dielectric measurement at various densities (Table 5.1). The theoretical density of Sigmacell is assumed to be 1.46 g/cm^3 [Sun05].

Table 6.2: Measured and theoretical dielectric properties calculated with two mixing rules for resin, hardener and fillers at 30°C

Material	Measurement		LR law with ς = -1		MG law	
	ε_r'	ε_r''	ε_r'	ε_r''	ε_r'	ε_r''
DGEBA	3.73	0.68				
Laromin	4.40	1.15				
Sigmacell	3.75	0.55				
D-LAR-S30	3.85	0.76	3.87	0.75	3.86	0.74
D-LAR-S50	3.83	0.73	3.86	0.72	3.84	0.70
D-LAR-S70	3.82	0.70	3.85	0.70	3.83	0.67

The calculated dielectric properties using the LR formula fit well with the measurement results. The maximal difference in dielectric constant and loss factor are only 0.08 (2.66%) and 0.02

(2.7%), respectively. The MG law also fits well with the measurement results with maximal differences of 0.13 (3.3%) and 0.04 (3.6%) for dielectric constant and loss factor, respectively. Based on the mixing rule at room temperature, the temperature dependent dielectric properties of mixtures can be derived from each constituent.

6.2 Temperature dependent model of resin and curing agents

Temperature dependent dielectric data can be modelled with many techniques. The suitable technique for modelling depends on the behavior of the material. The model that is commonly preferred to describe the dielectric behavior of non-conductive polar liquids in the microwave range is the Debye model. The formula of the Debye model has been already described in Chapter 2. A good explanation of the experimental results is frequently achieved with a distribution of relaxation times. Extension of the Debye model can be found in [Man98] where the model is enhanced by consideríng the specific behavior of two-dipole polarization. In the present study, as already seen in the measurements on each constituent, such kind of behavior has not been found. The measurements on DGEBA and PEA give evidence, that only the simple phenomenon of dipole polarization is necessary to describe the experimental observations.

6.2.1 DGEBA Model

Temperature dependent modelling of each constituent of mixtures is needed for the full modelling of curing. The behavior of the dielectric constant and loss factor of DGEBA indicates dipole polarization that can be described by the Debye model. In order to obtain the Debye model of DGEBA, the parameters of the Debye model were fitted to the measured dielectric properties employing minimizing of errors.

The Debye parameters obtained by fitting the correponding model from equations 2.11 and 2.13 to the measured dielectric constant of DGEBA are presented in Table 6.3. The dielectric constant fits well with RMSE 0.0208 and R-Square 0.9992. The parameters obtained for the dielectric constant can also be used in equation 2.12 to model the loss factor. Again, the measurement fits quite well with the model (RMSE = 0.05219) as shown in Table 6.3.

Table 6.3: Debye model parameters for the dielectric properties of DGEBA

Model parameter	Value	Fitting procedure	Value
ε_s	7.0 ± 0.01	RMSE of ε_r'	0.0208
ε_∞	3.6 ± 0.01	R^2 of ε_r'	0.9992
τ_0 (E-13 s)	9.28 ± 0.001	RMSE of ε_r''	0.0522
ζ (E-20 kJ/mol)	3.65 ± 0.01	R^2 of ε_r''	0.9773

The temperature dependent measured dielectric constant and the corresponding Debye model value are presented in Fig. 6.1. While the comparison of the measured loss factor and the result of the Debye model is shown in Fig. 6.2. The loss factor increases with increasing temperature and starts to decrease at approximately 100°C as an effect of dipolar relaxation. The comparison in Fig. 6.1 and 6.2 shows that both dielectric constant and loss factor of the model fit well with the measurement result.

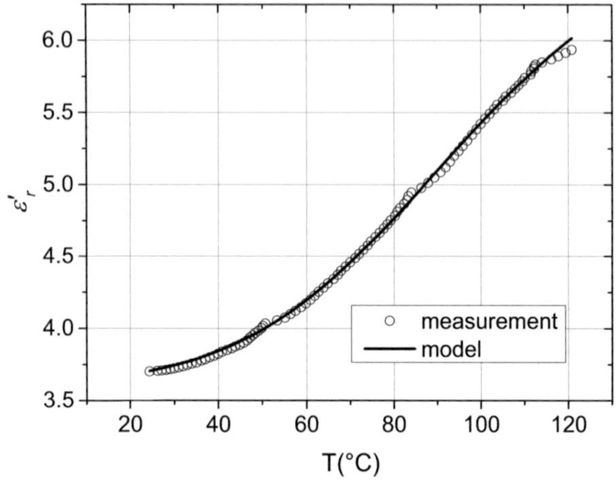

Figure 6.1: Debye model of the temperature dependent dielectric constant of DGEBA and comparison with measurement

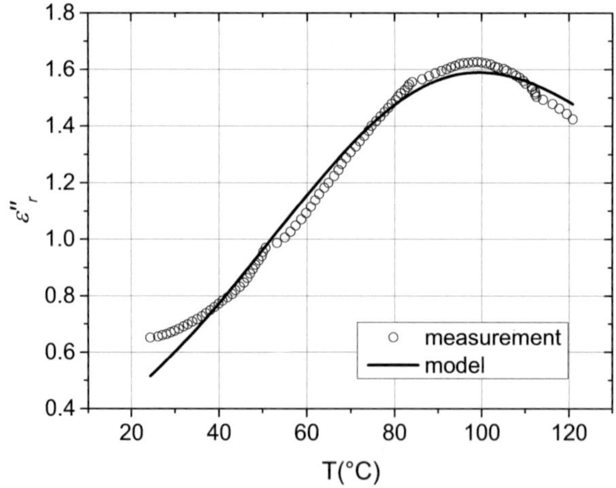

Figure 6.2: Debye model of the temperature dependent dielectric loss factor of DGEBA and comparison with experiment

6.2.2 PEA Model

Temperature dependent dielectric properties of PEA are also well represented by the Debye model. The model parameters and fitting procedure are summarized in Table 6.4. The dielectric constant of the Debye model fits well with the measured value as indicated by the small RMSE = 0.0735. The Debye parameters of the dielectric constant were used to investigate the fitting procedure for the measured loss factor. The Debye model also fits very well with the measured value of loss factor (RMSE = 0.0069).

Table 6.4: Debye model parameters for dielectric constant of PEA

Model parameter	Value	Fitting procedure	Value
ε_s	5.9 ± 0.01	RMSE of ε_r'	0.0735
ε_∞	3.1 ± 0.01	R^2 of ε_r'	0.9453
τ_0(E-13 s)	7.05 ± 0.001	RMSE of ε_r''	0.0069
ζ (E-20 kJ/mol)	2.26 ± 0.01	R^2 of ε_r''	0.9977

The result of the Debye model and the measured dielectric constant of PEA are shown in Fig. 6.3. The measured loss factorof PEA and the result of the Debye model are plotted in Fig. 6.4. The results shown in Fig. 6.3 and 6.4 clearly reveal that both dielectric constant and loss factor fit well with the measured values. Compared to DGEBA, the relaxation time of PEA is smaller as can be seen from Table 6.3 and 6.4 so that the peak of the loss factor is at the lower temperature (see Fig 6.2 and 6.4).

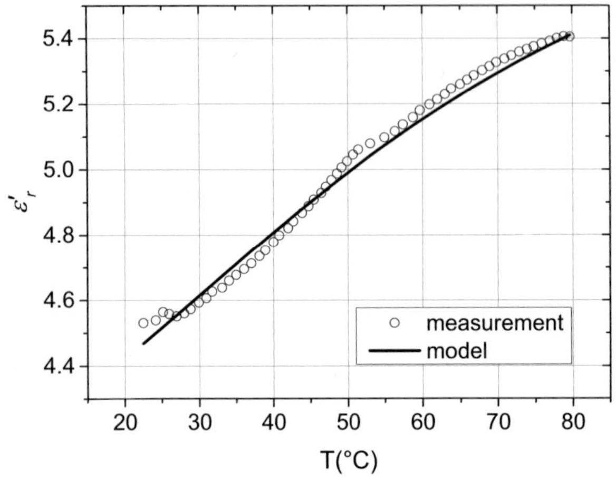

Figure 6.3: Debye model of temperature dependent dielectric constant of PEA and comparison with experiment

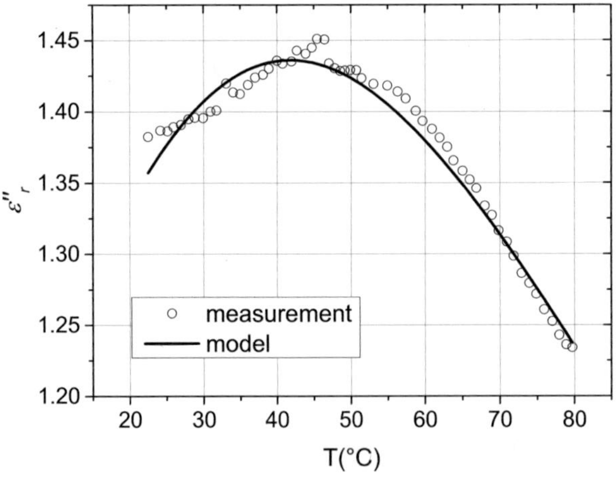

Figure 6.4: Debye model of temperature dependent dielectric loss factor of PEA and comparison with experiment

6.2.3 IPDA Model

Only the hardener IPDA could not be represented by the Debye model; instead, a modified sigmoidal model has been selected for temperature dependency. The sigmoidal had been used for modelling of temperature dependent dielectric properties in [JB11]. In the present work, the model of the temperature dependence of IPDA is given in the following formula:

$$\varepsilon_r{}' = \frac{a_{11}}{a_{21}+\exp(\ a_{31}+(T-a_{41}))} + a_{51} \tag{6.1}$$

$$\varepsilon_r{}'' = \frac{a_{12}}{a_{22}+\exp(\ a_{32}+(T-a_{42}))} + a_{52} \tag{6.2}$$

The corresponding model parameters for dielectric constant and loss factor are shown in Table 6.5 and 6.6, respectively.

Table 6.5: Sigmoidal model parameters for the dielectric constant of IPDA

Model parameter	Value	Fitting procedure	Value
a_{11}	1.35 ± 0.001	RMSE	0.0023
a_{21}	3.58 ± 0.002	R^2	0.9966
a_{31}	0.108 ± 0.0002		
a_{41}	50 ± 0.01		
a_{51}	3.99 ± 0.001		

The model fits well to the measured dielectric properties indicated by RMSE values of 0.0023 and 0.0043 for dielectric constant and loss factor, respectively.

Table 6.6: Sigmoidal model parameters for the loss factor of IPDA

Model parameter	Value	Fitting procedure	Value
a_{12}	0.617 ± 0.0003	RMSE	0.0043
a_{22}	1.85 ± 0.004	R^2	0.9760
a_{32}	0.0088 ± 0.00003		
a_{42}	50 ± 0.01		
a_{52}	0.571 ± 0.0002		

The measured dielectric constant and loss factor and the corresponding model are plotted in Fig. 6.5 and 6.6, respectively.

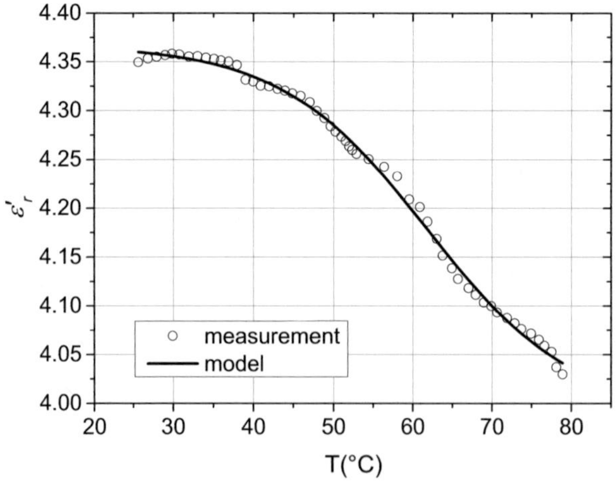

Figure 6.5: Sigmoid model of the temperature dependent dielectric constant of IPDA and comparison with experiment

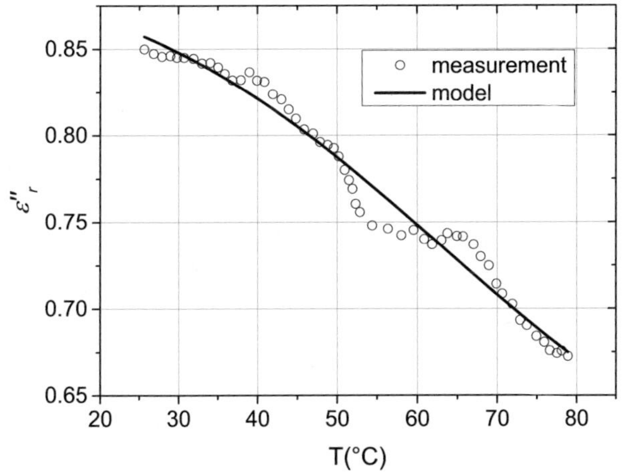

Figure 6.6: Sigmoid model of the temperature dependent dielectric loss of IPDA and comparison with experiment

Fig 6.6 shows small fluctuations in the loss factor which are within the uncertainty error of ± 5.2%. The above results show that the temperature dependent dielectric measurements of all constituent have good agreement with theoretical models which thus can be used for the modelling of curing.

6.3 Modelling of curing

In literature, the status of curing was monitored using the loss factor at low frequency up to 10 MHz. The curing process was investigated from the start of curing that can be identified by a decrease of the loss factor [BES93, HJPK13]. The curing degree was calculated with the following formula [HJPK13]:

$$\alpha(t) = \frac{\log(\varepsilon''_{t0}) - \log(\varepsilon''_{t})}{\log(\varepsilon''_{t0}) - \log(\varepsilon''_{\infty})} \tag{6.3}$$

ε''_{t0} is the loss factor at the beginning of curing and ε''_{∞} is the loss factor for infinite curing time. In experimental work, the loss factor for infinite time of curing can be taken at the time when the loss factor is close to its asymptotic value. This model only can be applied for curing processes leading to fully cured final products. For curing processes that do not produce a fully cured product, the loss factor of the fully cured material has to be estimated. As stated in [HJPK13], the above formula for calculation of the curing degree is only valid for isothermal temperature starting from time zero. This is an ideal condition which never can be achieved in real thermal curing.

The development of a novel model which is much closer to real curing is performed in the present work as already discussed in Chapter 5. The process temperature increases slowly up to a specific value (curing temperature). Different curing temperatures are investigated to describe the influence of temperature on the curing. The curing degree $\alpha(t)$ theoretically can be calculated by continuous integration over time. The proposed monitoring of curing is applied to the discrete time of measurement and the curing degree can be derived from the following differential formula:

$$\frac{\partial \alpha}{\partial t} = K(T) f(\alpha) \tag{6.4}$$

Time discrete measurements lead to the time derivative:

$$\Delta\alpha = \Delta t K(T) f(\alpha) \tag{6.5}$$

where $f(\alpha)$ is associated with the reaction model. The reaction model depends on the curing agent [Ehr10]. The variable $K(T)$ is the temperature dependent rate that can be commonly expressed by the Arrhenius formula [HJPK13]:

$$K(T) = A\,exp\left(\frac{-E_\alpha}{RT}\right) \tag{6.6}$$

where R = 8.3144621 J/(mol K) is the universal gas constant, A is a pre-exponential constant and E_α is the activation energy. The activation energy is the minimum energy needed for chemical system to start a chemical reaction. The investigated phenomenological reaction models in this study are the (n) order autocatalysis and Prout-Tompkins autocatalytic models [WMS02, HJPK13]. The (n) order autocatalysis model is written by the following formula:

$$f(\alpha) = (1-\alpha)^n(1+K_{cat}\alpha) \tag{6.7}$$

where the model parameters are α, n, and K_{cat}. Whereas the Prout Tompkins model is described by this formula:

$$f(\alpha) = (1-\alpha)^n\alpha^m \tag{6.8}$$

Where the model parameters are the degree of curing α, n and m. The calculation of a discrete curing degree for the Prout-Tompkins kinetic model is described by:

$$\alpha_p = \alpha_{p-1} + (t_p - t_{p-1})K(T)(1-\alpha)^n\alpha^m \tag{6.9}$$

The dielectric constant during curing can be calculated from the formula:

$$\varepsilon'_{rc} = \frac{\varepsilon'_{rm}}{\left(\frac{\varepsilon'_{rm}}{\varepsilon'_{\infty}}\right)^{\alpha}} \tag{6.10}$$

where the dielectric constant of the mixture ε'_{rm} depends on the percentages and behavior of each constituent. As discussed previously, the LR mixing rule formula is used to obtain the dielectric constant of the mixture.

The loss factor ε''_{rc} during curing can be calculated from the same type of formula given by:

$$\varepsilon''_{rc} = \frac{\varepsilon''_{rm}}{\left(\frac{\varepsilon''_{rm}}{\varepsilon''_{\infty}}\right)^{\alpha}} \tag{6.11}$$

The optimal model parameters are obtained by an optimization procedure. The optimization routine is used to minimize the RMSE of the measured and calculated dielectric properties. Based on the measurement results of chapter 5, both the dielectric constant and the loss factor represent the temperature and time dependent curing behavior. Therefore, the optimization routine uses both dielectric constant and loss factor as dependent variables. The dielectric constant and loss factor are optimized in a complex form so that the goal of the optimization is to minimize the RMSE of both parameters simultaneously. This method has a clear advantage compared to other previous publications that only use the loss factor to obtain the status of curing. The advantage of the present method is that the error in one parameter can be easily checked with the other parameter.

6.3.1 Modelling of curing with PEA

The temperature dependent dielectric properties of the mixture considered here are obtained from each component using the mixing rule in equation 2.14 with the LR exponential factor ς= -1. The mixing formulas for the dielectric properties of the D-PEA mixture are:

$$\varepsilon'_{rm} = (V{\varepsilon'_{rD}}^{\varsigma} + (1-V){\varepsilon'_{rP}}^{\varsigma})^{1/\varsigma} \tag{6.12}$$

$$\varepsilon''_{rm} = (V{\varepsilon''_{rD}}^{\varsigma} + (1-V){\varepsilon''_{rP}}^{\varsigma})^{1/\varsigma} \tag{6.13}$$

where V is the percentage of DGEBA in the mixture. The dielectric constants of the mixture, pure DGEBA and pure PEA are ε'_{rm}, ε'_{rD} and ε'_{rP} , respectively, and ε''_{rm}, ε''_{rD} and ε''_{rP} are the corresponding loss factors.

The kinetic reaction used in the model is the Prout-Tompkins autocatalytic model because the (n) order autocatalysis reaction model is not able to follow the fast exponential decay behavior of this mixture. The kinetic parameters obtained by minimizing the RMSE of both dielectric constant and loss factor of the model and the measurement results are summarized in Table 6.7.

Table 6.7: Kinetic parameters of D-PEA

Model parameter	Value	Fitting procedure	Value
E_α	51.69 ± 0.01 kJ/mol	RMSE	0.0570
log(A)	5.02 ± 0.0006 (s^{-1})	R^2	0.9946
n	1.61 ± 0.01		
m	0.81 ± 0.001		

The kinetic parameters in the Table 6.7 are comparable with the kinetic parameters of DGEBA with modified aliphatic amine pulished in [HJPK13]. There the kinetic parameters are E_α = 50.58 kJ/mol, log(A) = 4.82 s^{-1}, n = 1.89 and m = 0.39. Only the m parameter shows a significant difference because in [HJPK13] the assumption that curing starts from time zero with constant temperature is used. The hardeners are also different in this study. The characterization of curing of DGEBA with PEA could not be found in any literature.

A comparison of the temperature and time dependence of the dielectric properties obtained from the model with the results of measurements is presented in Fig. 6.7. The model of dielectric constant and loss factor fits quite well with the measurement results. Both dielectric constant and loss factor indicate the process of curing. In the first 50 minutes, they both increase because of the heating process from room temperature to the 80°C curing temperature. Going from room temperature up to approximately 73°C, the dielectric constant and the loss factor increase from 3.8 to 4.6 and from 0.8 to 1.4, respectively. From this time, both dielectric constant and loss factor start to decrease which indicates the process moves to a fast curing.

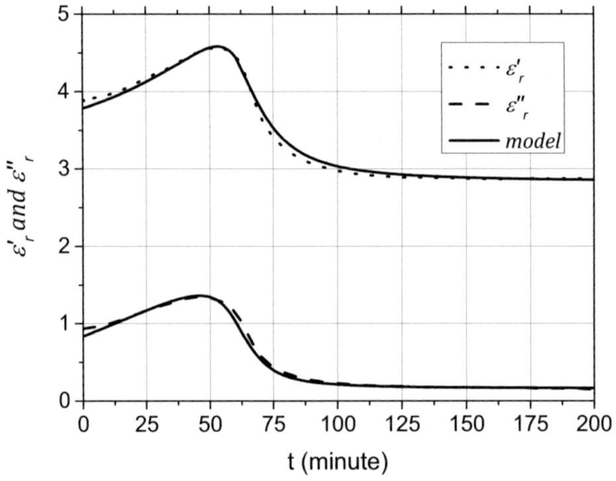

Figure 6.7: Modeled and measured time-dependent dielectric properties of D-PEA at 2.45 GHz during 80°C curing

The developed model also has been tested for curing at the different curing temperatures 70°C and 60°C.

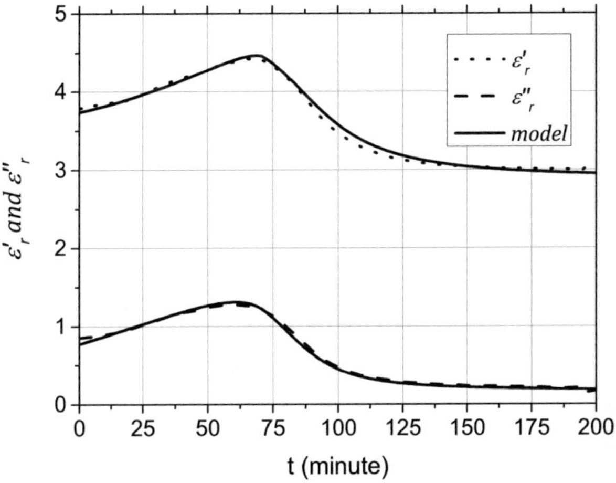

Figure 6.8: Predicted and measured time dependence of dielectric properties of D-PEA at 2.45 GHz during 70° C curing

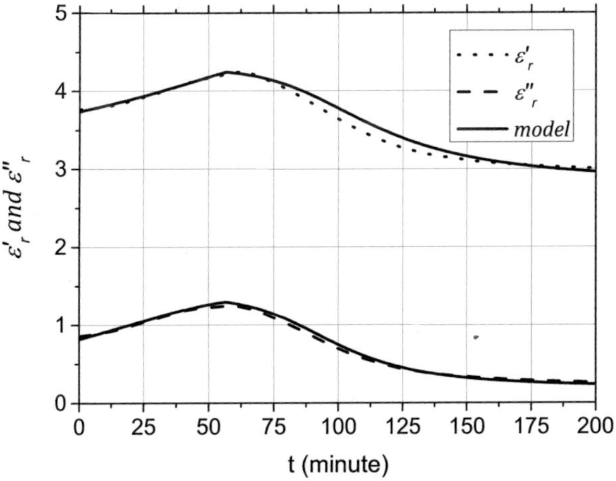

Figure 6.9: Predicted and measured time dependence of dielectric properties of D-PEA at 2.45 GHz during 60°C curing

The kinetic parameters of the 80°C process have been directly implemented to calculate the model at 70°C and 60°C without any optimization procedure. As previously explained, a correct model should have the same kinetic parameters at any different temperature of curing.

The measured and predicted dielectric properties during curing at 70°C and 60°C are presented in the Fig. 6.8 and 6.9, respectively. The measurements agree quite well with the dielectric properties of the model.

In order to verify the model, the final calculated degrees of curing are compared with those obtained from the DSC measurements (Table 6.8). The final product of D-PEA with 80°C curing temperature shows 100% and 99.41% curing degree, resulting from the DSC measurements and the model, respectively. So the difference between model and DSC measurement is around 0.6%.

Table 6.8: Curing degree of D-PEA

Curing temperature (°C)	Proposed model (%)	DSC measurement (%)
60	95.85	91.75
70	98.57	97.01
80	99.41	100

The advantage of the present model is that the prediction of the curing behavior can be used for different temperature schemes that are close to a real curing process. The comparison of the measurement results and the model shows that the model can be implemented at any different temperature. The validation with

DSC measurements shows that the model has quite good accuracy with a maximal difference of 4%.

The curing degree of D-PEA at different curing temperatures is shown in Fig. 6.10. The 80°C curing temperature scheme shows that the curing starts very slowly until 50 minutes (73°C) and then goes rapidly during the time from 50 to 100 minutes. After 150 minutes, the curing degree reaches 98%.

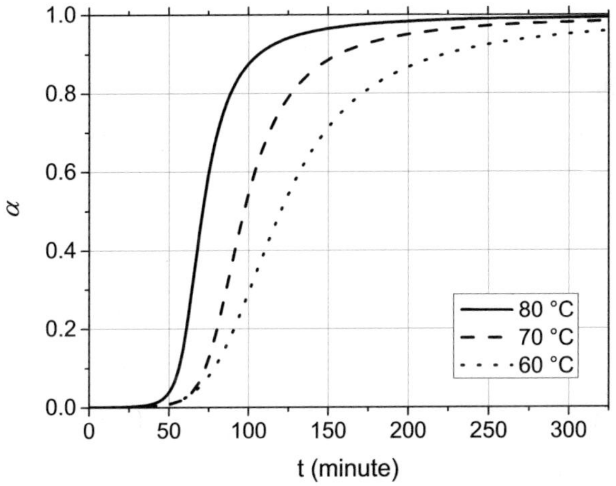

Figure 6.10: Curing degree of D-PEA at different temperatures obtained using 2.45 GHz dielectric monitoring

The curing degree at different temperatures shows that the higher the temperature, the faster the curing process. However, in real curing processes, the temperature and time scheme of curing should consider also the shrinkage and subsequent stress in the polymer composites so that stepwise heating is preferable [Ehr10]. The curing degree with stepwise heating can be also predicted with the model derived in the present work.

6.3.2 Modelling of curing with IPDA

Curing with different hardeners leads to different kinetic reactions. In modelling the curing of the mixture D-IPD, both the Prout-Tompkins autocatalytic and the (n) order autocatalysis models are not able to follow the behavior of the mixture. Therefore a modified kinetic reaction model has been implemented here by combining these two reaction kinetics:

$$f(\alpha) = (1-\alpha)^n \alpha^m (1 + K_{cat}\alpha) \tag{6.14}$$

The temperature dependent dielectric properties of the mixture D-IPD are obtained by applying the mixing rule formulas of equation 2.14.

$$\varepsilon'_{rm} = (V\varepsilon'^{\varsigma}_{rD} + (1-V)\varepsilon'^{\varsigma}_{rI})^{1/\varsigma} \tag{6.15}$$

$$\varepsilon''_{rm} = (V\varepsilon''^{\varsigma}_{rD} + (1-V)\varepsilon''^{\varsigma}_{rI})^{1/\varsigma} \tag{6.16}$$

where ε'_{rI} and ε''_{rI} are the dielectric constant and loss factor of IPDA, respectively. The obtained kinetic parameters of the mixture D-IPD are shown in Table 6.9.

Table 6.9: Kinetic parameters of D-IPD

Fitting parameter	Value	Fitting procedure	Value
E_α (kJ/mol)	48.63 ± 0.0350	RMSE	0.0712
log(A) (s^{-1})	5.31 ± 0.0001	R-Squared	0.9576
n	2.32 ± 0.0020		0.9573
m	1.08 ± 0.0001		
K_{cat}	-0.98 ± 0.0001		

The kinetic parameters of the mixture D-IPDA could not be found in any other literature. Compared with the model of D-PEA, the activation energy of D-IPD is slightly lower 2.02 kJ/mol. This confirms that IPDA is more reactive compared to the PEA as shown in the datasheet in the Tables 4.9 and 4.10.

The comparison of the model and the measurement of curing of DGEBA and IPDA at 80°C is shown in Fig. 6.11.

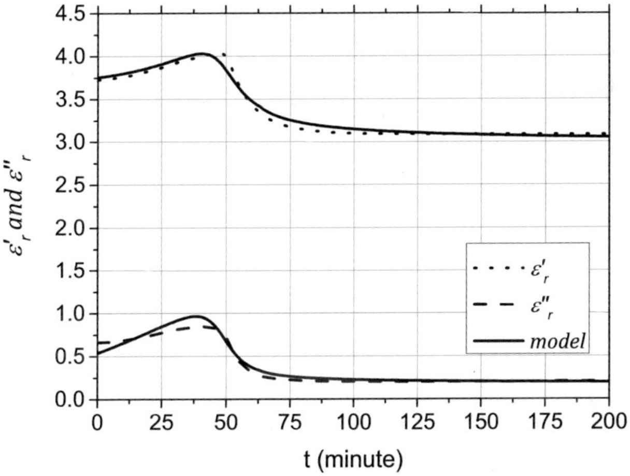

Figure 6.11: Modeled and measured dielectric properties of DGEBA-IPD at 2.45 GHz during 80 °C curing

Verification of the model also has been done for the D-IPD mixture at different curing temperatures by comparing the calculated curing degree of the model with DSC measurements as shown in Table 6.10. The final product of D-IPD with 80°C curing temperature shows 93.35% and 89.53% curing degree for the DSC measurements and the model, respectively. So the difference of calculated and measured curing degree is approximately 4%.

Table 6.10: Final curing degree of D-IPD

Curing temperature (°C)	Present model (%)	DSC measurement (%)
60	88.58	83.59
70	91.28	89.24
80	93.22	89.53

The curing degree of the final products obtained at 70°C and 60°C curing temperature also have been measured with the DSC method and compared with the values calculated using the developed model as shown in Table 6.10. The results show that the maximal uncertainty is around 5.0% for the final product with 60°C curing temperature. This accuracy is less compared to the model of D-PEA which is as expected because IPDA is the fastest harderner compared to the two others. A fast curing process increases the uncertainty of the measured results. However, the accuracy is still below the maximum error of the dielectric measurement which is around 5.2%.

For all three different curing temperatures the measured and modeled curing degrees are approximately the same and relatively far away from 100%. This fact shows that these temperatures are not sufficient for triggering further cross-linking reactions. Post curing is needed to achieve a fully cured product. The influence of the curing temperature on the curing behavior can also be predicted with the developed model. The obtained kinetic parameters are simply used for the prediction by changing the temperature scheme. The predictions of the model are compared with the results of measurements in the Fig. 6.12 and 6.13 for 70°C and 60°C curing temperature, respectively.

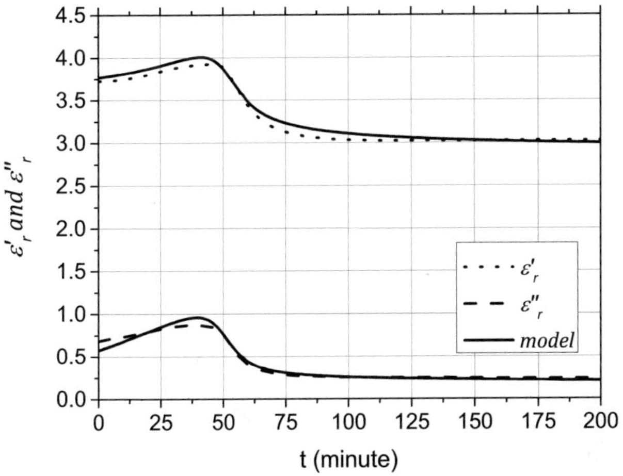

Figure 6.12: Predicted and measured dielectric properties of D-IPD at 2.45 GHz during 70°C curing

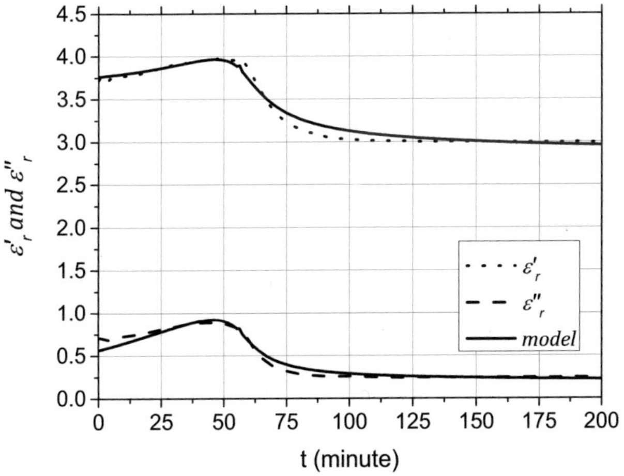

Figure 6.13: Predicted and measured dielectric properties of D-IPD at 2.45 GHz during 60°C curing

The predicted and measured dielectric properties show good agreement within the experimental uncertainties. The degrees of cure of D-IPD at the three different temperatures are shown in Fig. 6.14.

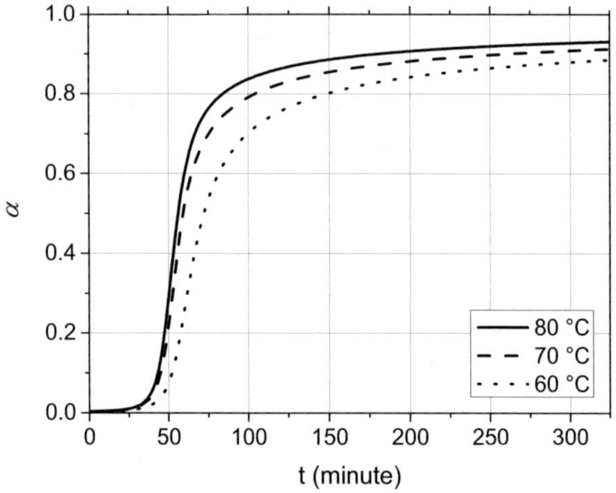

Figure 6.14: Cure degree of D-IPD at different temperatures obtained using 2.45 GHz dielectric monitoring

6.3.3 Modelling of curing with Laromin

Laromin is a mixture of PEA and IPDA. The measurement in Chapter 5 shows that Laromin has the intermediate characteristic of its constituent. The model for D-LAR cannot simply be derived from the models of D-PEA and D-IPDA. However, the temperature dependent dielectric properties of D-LAR can be directly obtained from the LR mixing rule of D-PEA and D-IPDA. The dielectric properties of the mixture are calculated using the following formulas:

$$\varepsilon'_{rm} = (V\varepsilon'_{rD}{}^{\varsigma} + V_I\varepsilon'_{rI}{}^{\varsigma} + V_P\varepsilon'_{rP}{}^{\varsigma})^{1/\varsigma} \tag{6.17}$$

$$\varepsilon''_{rm} = (V\varepsilon''_{rD}{}^{\varsigma} + V_I\varepsilon''_{rI}{}^{\varsigma} + V_P\varepsilon''_{rP}{}^{\varsigma})^{1/\varsigma} \tag{6.18}$$

where V_I and V_P are the volume fractions of IPDA and PEA, respectively. The formula 6.14 is used as kinetic reaction model. The kinetic parameters of D-LAR have been obtained by minimizing the error between measured and calculated dielectric properties for the 80°C curing temperature. The kinetic parameters are summarized in Table 6.11.

Table 6.11 : Kinetic parameters of D-LAR

Model parameter	Value	Fitting procedure	Value
E_α (kJ/mol)	51.71 ± 0.0400	RMSE	0.0712
log(A) (s^{-1})	5.36 ± 0.0001	R-Squared	0.9850
n	1.48 ± 0.0025		
m	0.93 ± 0.0001		
K_{cat}	-0.98 ± 0.0001		

Verification of the model is done by comparison of the curing degree of the final product after 80°C curing with the results of DSC measurements. The comparison shows that the model has an error of less than 2%.

Measured and modelled dielectric properties of D-Lar at 2.45 GHz during curing are presented in Fig. 6.15.

Table 6.12 : Final curing degree of D-LAR

Curing temperature (°C)	Present model (%)	DSC measurement (%)
60	91.11	89.45
70	94.53	92.54
80	96.54	95.80

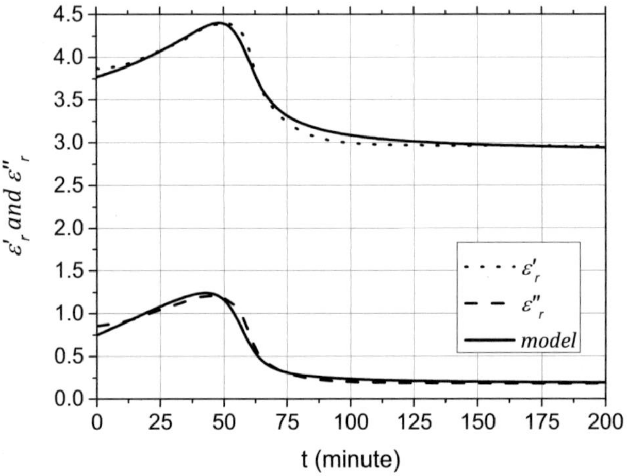

Figure 6.15: Modeled and measured dielectric properties of D-LAR at 2.45 GHz during 80°C curing

Fig. 6.15 shows that the dielectric properties of the model fit very well with the results of measurements. Measurements of curing of D-LAR at 70°C and 60°C have also been done. The measured dielectric properties are compared with the model as plotted in Fig. 6.16 and 6.17.

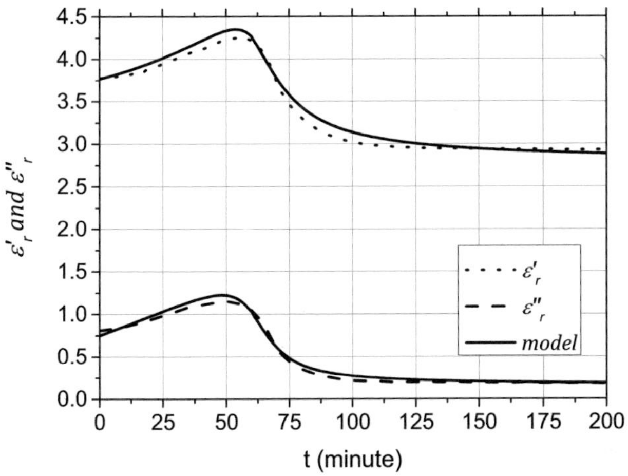

Figure 6.16: Predicted and measured dielectric properties of D-LAR at 2.45 GHz during 70°C curing

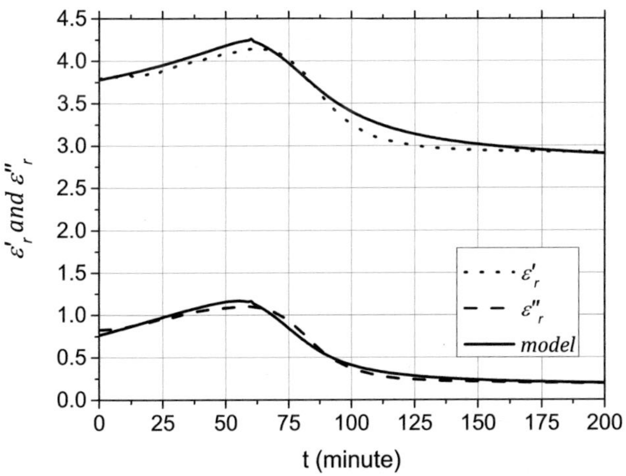

Figure 6.17: Predicted and measured dielectric properties of D-LAR at 2.45 GHz during 60°C curing

The degrees of curing of D-LAR for different curing temperatures are shown in Fig. 6.18. As in the D-PEA model, the curing degree of D-LAR depends on the curing temperature. The higher the temperature is, the faster the curing degree reaches the asymptotic value. This means that the time needed for the curing process is shorter at higher curing temperature.

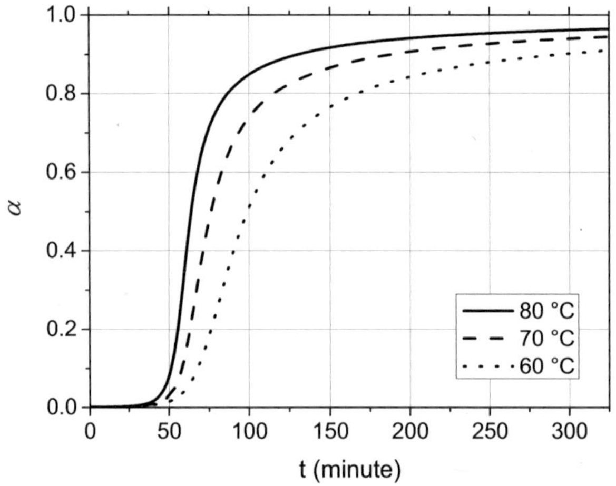

Figure 6.18: Curing degree of D-LAR at different curing temperatures obtained using 2.45 GHz dielectric monitoring

All results of D-LAR in these experiments do not reach 100% curing degree. So the curing temperature is not sufficient for obtaining a full curing process. This is confirmed by the T_g of Laromin which is around 85-94°C.

6.3.4 Modelling of curing with filler

The influence of the filler on the curing is investigated by measurements of mixtures with different weight percentages. The temperature dependent dielectric properties of the mixture are obtained by the following mixing formulas:

$$\varepsilon'_{rm} = (V\varepsilon'_{rD}{}^{\varsigma} + V_I\varepsilon'_{rI}{}^{\varsigma} + V_P\varepsilon'_{rP}{}^{\varsigma} + V_F\varepsilon'_{rF}{}^{\varsigma})^{1/\varsigma} \quad (6.19)$$

$$\varepsilon''_{rm} = (V\varepsilon''_{rD}{}^{\varsigma} + V_I\varepsilon''_{rI}{}^{\varsigma} + V_P\varepsilon''_{rP}{}^{\varsigma} + V_F\varepsilon''_{rF}{}^{\varsigma})^{1/\varsigma} \quad (6.20)$$

where V_F is the volume fraction of the filler and ε'_{rF} and ε''_{rF} are the dielectric constant and loss factor of the filler, respectively.

The kinetic model is selected based on the hardener which is used in the experiment. In this example, the model parameters are obtained by minimizing the means square error of the measured and calculated dielectric properties of D-LAR-S70. As in the model of D-LAR, the kinetic model in formula 6.14 is used. The kinetic parameters are summarized in Table 6.13.

Table 6.13: Kinetic parameters of D-LAR-S70

Model parameter	Value	Fiting procedure	Value
E_α (kJ/mol)	51.11 ± 0.030	RMSE	0.0434
log(A) (s^{-1})	5.30 ± 0.001	R-Squared	0.9923
n	1.71 ± 0.001		
m	1.03 ± 0.008		
K_{cat}	-0.98 ± 0.001		

The dielectric properties calculated by the model are compared with the measurements in Fig. 6.19. The result shows that the measured and modelled dielectric properties are in very good agreement.

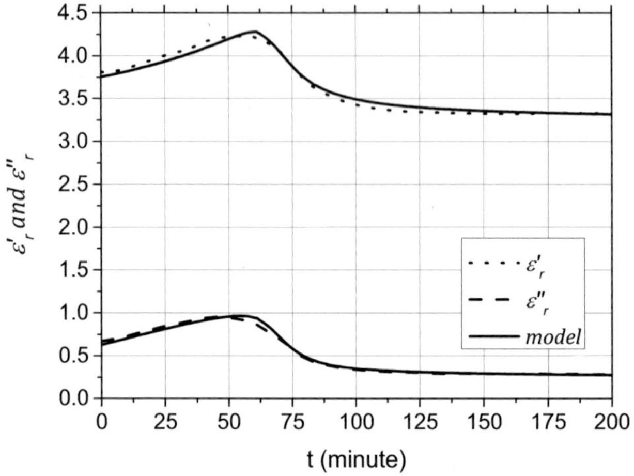

Figure 6.19: Comparison of modeled and measured dielectric properties of D-LAR-S70 at 2.45 GHz during 80°C curing

In order to verify the model, the final degrees of cure have been also measured with the DSC method. The comparison of calculations and measurements is given in Table 6.14. The result shows a maximum difference of approximately 2% in the degree of curing.

Table 6.14 : Final curing degree of D-LAR after curing at 80°C with different amount of filler

Mixture	Present model (%)	DSC measurement (%)
D-LAR-S70	91.85	93.53
D-LAR-S50	92.49	94.22
D-LAR-S30	93.39	94.51

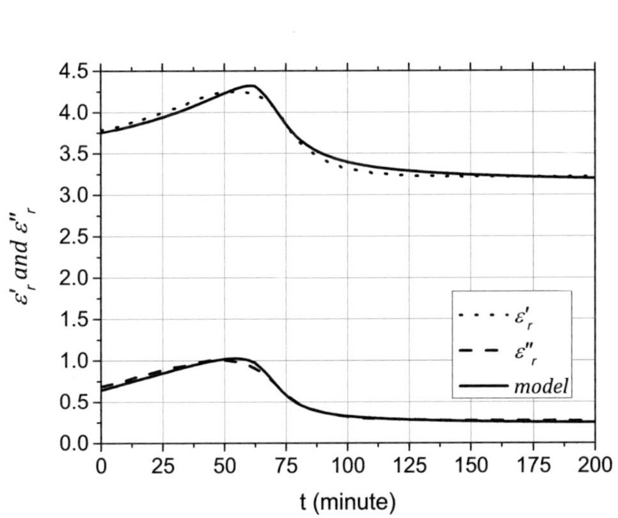

Figure 6.20: Comparison of modeled and measured dielectric properties of D-LAR-S50 at 2.45 GHz during 80°C curing

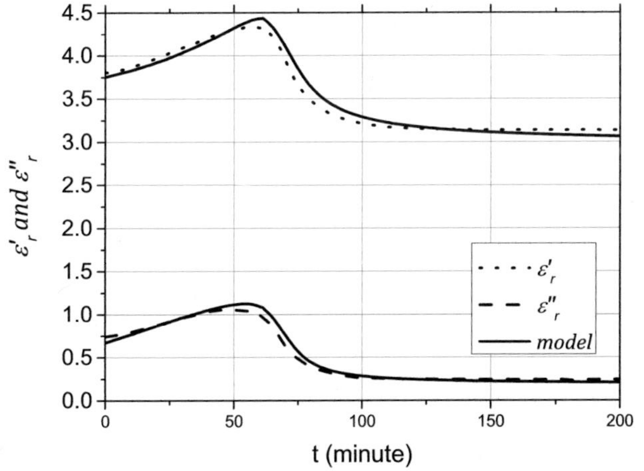

Figure 6.21: Comparison of modeled and measured dielectric properties of D-LAR-S70 at 2.45 GHz during 80°C curing

The model also compared to the measurements with different amount of filler which are 50 phr and 30 phr, respectively (Fig. 6.20 and 6.21). The obtained kinetic parameters in Table 6.15 are used for the model. The measured dielectric properties of D-LAR-S50 are compared with the results of the model in Fig. 6.20. Both dielectric constant and loss factor of the model match very well with the measurement result. The results of measurements and model for DGEBA-PEA-S30 also show good agreement as can be seen in Fig. 6.21.

The time dependence of the curing degree of D-LAR-with different amount of filler is plotted in Fig. 6.22.

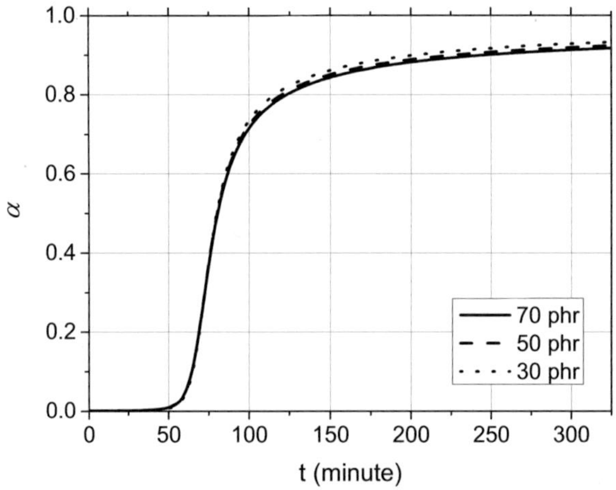

Figure 6.22: Curing degree of D-LAR-S70, D-LAR-S50 and D-LAR-S30 at 80°C obtained using 2.45 GHz dielectric monitoring

The result shows that the influence of the filler on the cure degree is small. The difference of the final curing degree between the experiments with the lowest and highest amount of filler is only 1.5%. Compared to the mixture without filler, the maximal cure degree is 4.5% lower (see Table 6.14 and 6.16). The filler makes the cross-linking reaction slower that depends on the percentage of the filler. This information gives proofs that the cellulose filler can widely used for the composite with predictable influence to the reaction.

7 Conclusion and outlook

The knowledge of the temperature dependent dielectric properties of materials to be processed is the prerequisite for the design of an appropriate microwave processing system and for simulation of related applications. In general, the lack of temperature dependent dielectric material data is the barrier for further development of microwave applications. Dielectric monitoring of curing at microwave frequencies with the non-resonant transmission-reflection method could not be found in any literature. The development of a method for curing monitoring at 2.45 GHz has the two advantages of obtaining temperature dependent dielectric property data in this microwave frequency range and of monitoring the status of curing.

In this work, a new system for monitoring of curing of polymer composites has been developed. A fully automatic system has been designed and implemented for temperature and time dependent dielectric measurements. The system is used for measurements on common resins, fillers and curing agents in large ranges of dielectric constant and loss factor of 1.0-10 and 0.1-10, respectively. The corresponding resolution limits of the system are 0.1 and 0.05 for dielectric constant and loss factor, respectively.

Measurements on DGEBA epoxy resin and PEA hardener show that the temperature dependent dielectric properties follow the Debye model. In contrast to this, the measurements with the hardener IPDA cannot be modelled with the Debye model; instead, a sigmoid function model of temperature dependency has been applied.

Investigations on curing with different independent parameters have been performed. Measurements on mixtures with different hardeners show that PEA, Laromin and IPDA have different speed of curing. Each hardener needs a certain temperature for reaching full curing. The PEA has slower curing rates followed by Laromin and IPDA, respectively. However, PEA has the lowest curing temperature. The T_g measurement for PEA shows that PEA can be used for applications below 78°C.

The measurements on the filler show that the filler reduces the speed of curing. The filler also increases curing temperatures of mixtures. The results show that the higher the percentage of filler is, the slower is the obtained curing rate. However, the curing degree can be increased by simply increasing the temperature. The experiments on the bio filler S3504 show that reducing the percentage of resin and hardener is possible by adding more filler and decreasing of the curing degree can be compensated by increasing the curing temperature.

A new curing model has been successfully developed. The novelty of this model is that it can be used for a more realistic description of the curing process. The model allows simulating the real curing scheme. It is able to predict the behavior of curing from room temperature up to a certain maximum temperature and to predict common schemes of curing with stepwise heating. This ability does not exist in other published models that usually assume that the curing temperature is constant from the beginning of the curing process (constant from time zero). The present model has been proven with DSC measurements which show that the maximal uncertainty of the model is around 5%.

In contrast to existing models, the new model uses both dielectric constant and loss factor for curing modelling. The new approach is more accurate than other published approaches which only use the loss factor for the modelling. The data from the present study can be used for designing appropriate polymer composites curing systems for the investigated mixtures.

In the future, the development of curing monitoring systems for other ranges of dielectric properties is needed. Based on the present results, the development is possible by designing appropriate crucibles for different materials. However, experimental investigation should be done to validate future developments.

Bibliography

[ADA09] H. T. Akbas, H. Darmus and G. Ahmetli. Dielectric properties of cured epoxy with teta. *Ozean Journal of Applied Sci.*, vol. 2, no. 4, pp. 443-449, 2009.

[AFT06] M. J. Akhtar, L. E. Feher and M. Thumm. A waveguide-based two-step approach for measuring complex permittivity tensor of uniaxial composite materials. *IEEE Trans. Microwave Theory Tech.*, vol. 54, no. 5, pp. 2011-2022, 2006.

[AFT07] M. J. Akhtar, L. E. Feher, and M. Thumm. A close form solution for reconstruction of permittivity of dielectric slabs placed at the center of a rectangular waveguide. *IEEE Geoscience and Remote Sensing Letter*, vol. 4, no. 1, pp. 122–126, 2006.

[AFT09] M. J. Akhtar, L. E. Feher, and M. Thumm. Noninvasive procedure for measuring the complex permittivity of resins, catalysts and other liquids using a partially filled rectangular waveguide structure. *IEEE Trans. Microwave Theory Tech.*, vol. 57, no. 2, pp. 458–470, 2009.

[AM08] K. J. Astrom and R. M. Murray. *Feedback Systems.* Princeton University Press, 2008.

[AT13] M. J. Akhtar and M. Thumm. Measurement of complex permittivity of cylindrical objects in the E-Plane of a rectangular waveguide. *IEEE Trans. Geoscience and Remote Sensing.* vol. 51, no.1, pp. 125–130, 2010.

[Bar78] B. A. Barry. *Errors in Practical Measurement inScience, Engineering and Technology*. John Wiley and Sons, 1978.

[Bas13a] BASF. *Technical Information of Baxxodur EC 301*. 2013.

[Bas13b] BASF. *Technical Information of Baxxodur EC 201*. 2013.

[BG99] A. K. Bledzki and J. Gassan. Composites reinforced with cellulose baded fibres. *Progress in Polymer Sci.*, vol. 24, no. 2, pp. 221-274, 1999.

[BSEB93] B. Binder, F.S. Schmid, P. Elsner and G. Busse. Simultaneous monitoring of viscoelastic and dielectric properties of curing epoxy resin using a vibrating electrode. *Colloid and Polym Sci.,* vol. 271. no.1, pp. 22-29, 1993.

[CCFD03] J. M. Catala-Civera, A. J. Canos, F. L. Penaranda-Foix, and E. R. Davo. Accurate determination of the complex permittivity of materials with transmission reflection measurements in partially filled rectangular waveguides. *IEEE Trans. Microwave Theory Tech.,* vol. 51, no. 1, pp. 16–24, 2003.

[Che] Chemspider Website- search and share chemistry. `http://www.chemspider.com/Molecular-Formula/C21H24O4.`

[Chu06] B. K. Chung. A convenient method for complex permittivity measurement of thin material at microwave frequency. *J. Phys. D Appl. Phys.,* vol. 36, pp. 1926–1931, 2006.

[Col91] R.E. Collin. *Field Theory of Guided Waves.* IEEE Press, 1991.

[CON+04] L. F. Chen, C. K. Ong, C. P. Neo V. V. Varadan and V. K. Varadan. *Microwave Electronics Measurement and Material Characterisation.* Wiley, 2004.

[Cou70] W. E. Courtney. Analysis and evaluation of a method of measuring the complex permitivitty and permeability of microwave insulators. *Trans. Microwave Theory Tech.,* vol. 18, no. 8, pp. 476-485, 1970.

[CP43] W. S. M. Culloch and W. H. Pitts. A logical calculus of ideas immanent in nervous activity. *Bulletin of Mathematical Biophysics,* vol. 5, pp. 115-133, 1943.

[Cst] CST – Computer Simulation Technology – Official Website. `http://www.cst.com`.

[DDRG77] J. G. Dorsey, G. F. Dorsey, A. C. Rutenberd and L. A. Green. Determination of the epoxy equivalent weight of Glycidyl Ether by proton magnetic resonance spectrometry. *Analytical Chemistry,* vol. 49 no. 8, pp. 1144-1145, 1977.

[Deb29] P. Debye. *Polar Molecules.* Chemical Catalog Co, 1929.

[DHC+99] B.L. Denq, Y.S. Hu, L.W. Chen, W.Y. Chiu and T.R. Wu. The curing reaction and physical properties of DGEBA/DETA epoxy resin blended with propyl ester phosphazene. *J. Appl Polymer Sci,* vol. 74, pp. 229-237, 1999.

[Ehr10] G.W. Ehrenstein. *Polymeric Materials: Structure - Properties - Application.* Hanser Gardner Publication, 2010.

[EMY07] E. E. Eves, E. K. Murphy, and V. V. Yakovlev. Practical aspects of complex permittivity reconstruction with neural-network-controlled FDTD modeling of a two-port fixture. *J. Microwave Power & Electromagnetic Energy,* vol. 41, no. 4, pp. 81-94, 2007.

[FT04] L. E. Feher and M. Thumm. Microwave innovation for industrial composite fabrication—The HEPHAISTOS Technology. *IEEE Trans. Plasma Sci.,* vol. 32, no. 1, pp. 73-79, Feb 2004.

[GLV00] A. V. Goncharenko, V. Z. Lozovski and E.F. Venger. Lichteneker's equation: applicability and limitations. *Optic Communication,* vol. 174, pp. 19-32, 2000.

[Has09] U. C. Hasar. Thickness-independent complex permittivity determination of partially filled thin dielectric material into rectangular waveguide. *Progress in Electromagnetic Research, PIER.,* vol. 93, pp. 189–203, 2009.

[Has10] U. C. Hasar. Accurate complex permittivity inversion from measurements of a sample partially filling a waveguide aperture. *IEEE Trans. Microwave Theory Tech.,* vol. 58, no. 2, pp. 451–457, 2010.

[Hay98] S. Haykin. *Neural Network: A Comprehensive Foundation.* Prentice Hall Int., 1998.

[HJPK13] R. Hardis, J. L. P. Jessop, F. E. Peters, M. R. Kessler. Cure kinetics characterization and monitoring of an epoxy resin using DSC, Raman spectroscopy, and DEA. *Composite, Elsevier, Part A,* vol. 49, pp. 100-108, 2013.

[HP97] Hewlett Packard. *HP 85070B Dielectric Probe Kit User Manual.* 1997.

[Jav90] J. Baker-Javis. *Transmission/reflection and Short Circuit Line Permittivity Measurements.* Technical Note, NIST Co., 1990.

[JB11] C. Ji and C. L. Brace. Expanded modelling of temperature dependent dielectric properties for microwave ablation. *J. Phys. Med. Biol.* vol. 56, pp. 5249-5264, 2011.

[JVK90] J. Baker-Javis, E. J Vanzura and W. A. Kissick. Improved Technique for Determining Complex Permittivity with the Transmission/Reflection Method. *Trans. Microwave Theory Tech.,* vol. 38 no. 8, pp. 1096-1103, 1990.

[KC99] H. Kim and K. Char. Dielectric Changes During the Curing of Epoxy Resin Based on the Diglycidyl Ether of Bisphenol A (DGEBA) with Diamine. *Bull. Korean Chem. Soc.,* vol. 20, no. 11, pp. 1329-1333, 1999

[KN92] A. W. Kraszewski and S. O. Nelson. Observations on cavity perturbation by dielectric objects. *IEEE Trans. Microwave Theory Tech.,* vol. 40 no.1, pp. 151-152, 1990.

[KW98] J. Krupka and C. Weil. Recent advances in metrology for electromagnetic characterization of material at microwave frequency. *Proc. 12th Int. Conf. Microwave Radar,* vol. 4, pp. 243-253, 1998.

[LFT99] G. Link, L. Feher and M. Thumm. Sintering of Advanced Ceramics Using a 30 GHz, 10 kW, CW

Industrial Gyrotron. *IEEE Trans. Plasma Sci.,* vol. 2, no. 2, pp. 547-554, 1999.

[Leu13] Leuna GmbH. *Datasheet of Epilox A 18-00.* 2013.

[Man98] P. Q. Mantas. Dielectric response of materials: extension to the Debye model. *J. European Ceramic Soc.,* vol. 19. pp. 2079-2086, 1999.

[Mat] Matlab – Matrix Laboratory – Official Website. `http://www.mathwork.com.`

[Lin+11] G. Link et al. Schlussbericht für das BMBF-Verbundprojekt Innovative, modulare Mikrowellentechnologie zur Herstellung von Faserverbundstrukturen Förderkennzeichen: 01RI05133, -135-140, -282 Laufzeit: 01.09.2006 – 31.05.2011;`http://www.cleanerproduction.de/fileadmin/assets/01RI05133_Abschlussbericht.pdf.`

[MKB92] E. Marrand, K. R. Baker and J. D. Graybeal. Comparison of reaction mechanisms of epoxy resins undergoing thermal and microwave in situ measurements of microwave dielectric properties and infrared spectroscopy. *Macromolecules,* vol. 25, pp. 2243-2252, 1992.

[NBFN03] L. Nunez, S. G. Barriero, C. A. G. Fernandez and M. R. Nunez. Use of dielectric analysis to complement previous thermoanalytical studies on the system diglycidyl ether of bisphenol A/1,2 diamine cyclohexane. *Polymer,* vol 45, pp. 1167-1175, 2004.

[NND04] A. Nesbitt, O. Novabpour, B. Degamber, C. Nightingale, T. Mann, G Fernando and R. J. Day.

Development of a microwave calorimeter for simultaneous thermal analysis, infrared spectroscopy and dielectric measurement. *Measurement Sci. Tech.*, vol 15, pp. 2313-2324, 2004.

[NR70] A. M. Nicolson and G. F. Ross. Measurement of the intrinsic properties of materials by time-domain techniques. *IEEE Trans. Instrumentation and Measurement*, vol. 19, no. 4, pp. 377–382, 1970.

[OPRT02] R. Omni, G. Pelosi, C. Riminesi and M. Todesco. Neural network for microwave characterization of material samples in rectangular cavities. *Microwave and Optical Tech. Letter*, vol. 35, no. 6, pp. 609-611, 2002.

[Par09] S. J. Park. Studies on cure behaviors, dielectric characteristics and mechanical properties of DGEBA/Poly (ethylene terephthalate) blends. *Macromolecular Research*, vol. 17, no. 8, pp. 585-590, 2009.

[PFSJ03] A. Penirshke, J. Freese, M. Schüßer and R. Jakoby. Neural network for microwave charazterization of material in rectangular cavities. *Proc. 3rd IEEE Int. Simp. Signal Processing and Information Technology*, pp. 609-611, 2003.

[Poz12] D. M. Pozar. *Microwave Engineering*. John Wiley and Sons, 2012.

[QB12] F. Qin and C. Brosseau. A review and analysis of microwave absoroption in polymer composites filled with carbonaceous particles, *J. Applied Physics*, vol. 111, 061301, pp. 01-24, 2012.

[QG96] P. Queffelec and P. Gelin. Influence of higher order mode on the measurements of complex permittivity using a microstrip discontinuity, *IEEE Trans. Microwave Theory and Techniques,* Vol. 44, No. 6, pp. 816-824, 1996.

[Qui] QuickWave 3D - QWED – Official Website. `http://www.qwed.com.pl/qw_3d.html`.

[RBF05] L.N. Rugueira, S.G. Barreiro and C.G. Fernandez. Study of influence of isomerism on the curing properties of the epoxy resin DGEBA (n=0)/1,2 DCH by DEA and MDSC, *J. Thermal Analysis and Calorimetry*, vol. 82, pp. 797-801, 2005.

[RSWZ07] H. Ren, J. Sun, B. Wu and Q. Zhou. Synthesis and curing properties of a novel novalac curing agent containing naphthyl and dicyclopendiene moieties Cin. *J. Chem. Eng.,* vol. 15 no.1 pp. 127-131, 2007.

[SC83] P. N. Sen and W. C. Chew, The frequency dependent dielectric and conductivity response of sedimentary rocks. *J. Microwave Power*, vol. 18, no.1, pp. 96-105, 1983.

[Sig14] Sigma-Aldrich. *Safety Data Sheet of S3504*. 2014.

[SHA+11] P.F Sung, Y.L Hsieh, K. Angones, D. Dunn, R.J King, R. Machbitz, A. Christianson, W.J. Chappel, L.S. Taylor and M.T. Harris. Complex dielectric properties of microcrystalline cellulose anhydrous lactose, and α-lactose monohydrate powders using a microwave based open-reflection resonator sensor. *J. Pharmaceutical Sci.,* vol. 100, no.7, pp. 2920-2934, 2011.

[Sih00] A. Sihvola. Mixing rules with complex dielectric coefficients. *Subsurface Sensing Technologies and Applications*, vol 1, no. 4, pp. 393-415, 2000.

[SM78] S. S. Stuchly and M. Matuszewski. A combined total reflection-transmission method in application to dielectric spectroscopy. *IEEE Trans. Instrumentation & Measurement.*, vol. 27, no. 3, pp. 285–288, 1978.

[SKL+14] Y. Sun, T. Kuehner, G. Link, T. Kayser and J. Jelonnek. Advanced temperature control of distributed microwave coupling systems. *Proc. 16th Seminar on Computer Modelling in Microwave Power Engineering*, pp. 63-67, 2014.

[SUC10] V.A. Scarlatache, M. Ursan and R. C. Ciobanu. Dielectric properties characterization of conductive polymeric materials at temperature and frequency variation. *Proc. 12th Int. Conf. Optimization of Electrical and Electronic Equipment, OPTIM*, pp. 208-213, 2010.

[Sun05] C. Sun. True densitx of microcrystalline cellulose. *J. Pharmaticeutical Sci.*, vol. 94. no. 10, pp. 2132-2134, 2005.

[TKAS12] X.Z. Tang, P. Kumar, S. Alavi and P. Sandeep. Recent advances in biopolymers and biopolymer-based nanocomposites for food packaging materials. *Critical Reviews in Food Sci. and Nutrition*, vol. 52, no. 5, pp 426-442, 2012.

[TOH+08] N. Tagami, M. Okada, N. Hiraj, Y. Ohki, T. Tanaka, T. Imai, M. Harada and M. Ochi. Dielectric properties of epoxy/clay nanocomposites – effects of curing agent and clay dispersion method. *IEEE. Trans.*

Dielectric and Electrical Insulation, vol. 15, no. 1, pp. 24-32, 2008.

[WMS02] S.R. White, P.T. Mather, and M.J. Smith. Characterization of the cure state of DGEBA-DDS epoxy using ultrasonic, dynamic mechanical and thermal probe. *Polymer Engineering and Sci.,* vol. 2, no. 1, pp 51-67, 2002.

[Wei74] W. B. Weir. Automatic measurement of complex dielectic constant and permeability at microwave frequency. *Proc. IEEE,* vol. 62, no. 1, pp. 33-36, 1974.

[Yar01] P.K. Yarlagadda. Study on microwave dielectric properties of epoxy resin mixtures used for rapid product development. *Acta Metallurgica Sinica (English Letters),* vol. 14 no.1, pp. 1-12, 2001.

[YKK09] G.Y Yun, J.H Kim and J. Kim, Dielectric and polarization behavior of cellulose electro-active paper (EAPap). *J. Phys S. Appl Phys.,* vol. 42, pp.1-6, 2009.

[YME05] V. V. Yakovlev, E. K. Murphy and E. E. Eves. Neural network for FDTD-backed permittivity reconstruction, *J. COMPEL,* vol. 24, no. 1, pp. 291-304, 2005.

[ZCKH04] L. Zong, G. Charvat, L. Kempel and M Hawley Automated Method for Characterizing Temperature Dependent Dielectric Materials. *Proc. 26 Ann. Meeting and Symposium AMTA Conf.,* pp. 380-390, 2004.

[ZLV98] T. Zakri, J. L. Laurent and M. Vauclin. Theoretical evidence for 'Licthenecker's mixture formulae' based on effective medium theory. *J. Phys. D. Appl.* vol. 31, pp 1589-1594, 1998.

List of own publication

[1] D. Prastiyanto, G. Link, U. Arnold, M. Thumm and J. Jelonnek. Time and temperature dependent dielectric measurement of thermosetting resins. In *Proc. 16th Seminar Computer Modeling in Microwave Power Engineering*, Karlsruhe, Germany, May 2014.

[2] D. Prastiyanto and G. Link. Temperature dependent dielectric Measurements at 2.45 GHz. In *Proc. Int. Engineering Conf. Unnes Conservation*, Semarang, Indonesia, Jan 2013.

[3] D. Prastiyanto, G. Link, M Thumm and J. Jelonnek Temperature dependent measurements of dielectric materials at 2.45 GHz. In *Proc. 1st Annual Meeting, KIT Center of Energy*, Karlsruhe, Germany, Jun 2012.

[4] D. Prastiyanto, G. Link and M Thumm. Dielectric measurements using transmission line method and different sample geometries in a WR340 standard waveguide. In *Proc. 13th Seminar Computer Modeling in Microwave Engineering & Applications*, Thun, Switzerland, Mar 2011.

[5] D. Prastiyanto, M. J. Akhtar, L. E. Feher, S. Stanculovic, V. Nuss, S. Layer and T. Seitz. Mechatronic temperature-dependent dielectric properties measurement at 2.45 GHz using a rectangular waveguide. In *Proc. 12th Int. Conf. on Microwave and High Frequency Heating (AMPERE)*, Karlsruhe, Germany, Sept 2009.

[6] D. Prastiyanto, L. E. Feher and M. Thumm. Development of dielectric properties measurement for industrial material using a rectangular waveguide. In *1st KIT PhD Symposium*, Karlsruhe, Germany 2009.

Acknowledgements

This work has been accomplished during my stay as research assistant at Institute for Pulsed-Power and Microwave Technology (IHM) of Karlsruhe Institute of Technology (KIT). First, I would like to express my deepest gratitude to Prof. Dr. rer. nat. Dr. h.c. Manfred Thumm for his excellent support, patience, motivation and guidance as my doctoral supervisor. I would like to thank also my co-supervisor Prof. Dr.-Ing. Ellen Ivers-Tiffée.

My best gratitude goes to Dr. Guido Link who always supported me tirelessly and who was always available for discussion on all research works that I did during my stay at the group "Material Process Technology with Microwaves". My thanks also go to all the group members who always supported me all the time. I would like to thank also Dr. Arnold Ulrich for our discussions about materials and the curing process, to Sigrid Silbernagel-Donath for helping me to prepare the samples and to Ruben Kleis as my student assistant.

I would like to deeply thank Prof. Dr.-Ing. John Jelonnek who always pushed me in a positive way and supported me to finish this work in his Institute. I would like to thank also all members of IHM for the excellent atmosphere needed to do successful research.

I would like to say thanks to my wife Erni Priliawati, my son Linove Syamsa Hezkia and my daughter Clearissa Prastiyanto for accompanying my stay in Germany and to my parents.

Karlsruhe, February 2016

Dhidik Prastiyanto

KARLSRUHER FORSCHUNGSBERICHTE AUS DEM
INSTITUT FÜR HOCHLEISTUNGSIMPULS- UND MIKROWELLENTECHNIK
(ISSN 2192-2764)

Band 1 **MATTHIAS BERINGER**
Design Studies towards a 4 MW 170 GHz Coaxial-Cavity Gyrotron. 2011
ISBN 978-3-86644-663-2

Band 2 **JENS FLAMM**
Diffraction and Scattering in Launchers of
Quasi-Optical Mode Converters for Gyrotrons. 2012
ISBN 978-3-86644-822-3

Band 3 **MATTIA DEL GIACCO**
Investigation of Fretting Wear of Cladding Materials in Liquid Lead. 2013
ISBN 978-3-86644-960-2

Band 4 **AMITAVO ROY CHOUDHURY**
Investigations of After Cavity Interaction in Gyrotrons
Including the Effect of Non-uniform Magnetic Field. 2013
ISBN 978-3-7315-0129-9

Band 5 **MICHAEL BETZ**
The CERN Resonant WISP Search (CROWS). 2014
ISBN 978-3-7315-0199-2

Band 6 **ANDREAS SCHLAICH**
Time-dependent spectrum analysis of high power gyrotrons. 2015
ISBN 978-3-7315-0375-0

Band 7 **DHIDIK PRASTIYANTO**
Temperature- and Time-Dependent Dielectric Measurements
and Modelling on Curing of Polymer Composites. 2016
ISBN 978-3-7315-0424-5